Revise A2 Chemistry for Salters (OCR)

Dave Newton, Alasdair Thorpe and Chris Otter

www.heinemann.co.uk

✓ Free online support
✓ Useful weblinks
✓ 24 hour online ordering

01865 888058

Heinemann is an imprint of Pearson Education Limited, a company
incorporated in England and Wales, having its registered office at
Edinburgh Gate, Harlow, Essex, CM20 2JE
Registered company number: 872828

Heinemann is a registered trademark of
Pearson Education Limited

© Dave Newton, Alasdair Thorpe, University of York Science Education Group 2004

First published 2004

09 08 07
10 9 8 7

British Library Cataloguing in Publication Data is available
from the British Library on request.

ISBN: 978 0 435583 47 7

Edited by Sara Hulse

Index compiled by Paul Nash

Designed and typeset by Saxon Graphics Ltd, Derby

Original illustrations © Pearson Education Limited, 2004

Printed and bound in Great Britain by Ashford Colour Press Ltd, Gosport, Hampshire

Acknowledgements
The examination questions and mark schemes are reproduced by kind permission of OCR and
are taken from the following papers:

Pages 80–87, Q1–Q6: January 2002
Page 87, Q7: June 2002
Page 89, Q8: January 2003
Page 90, Q9: January 2002

Every effort has been made to contact copyright holders of material reproduced in this book.
Any omissions will be rectified in subsequent printings if notice is given to the publishers.

Contents

How to use this revision guide

This revision guide is for the OCR Chemistry (Salters) A2 course. You may take examinations in January or June, or just in June at the end of the course. This guide covers the two written examinations for the A2 course, **Module 2849, Chemistry of Materials** and **Module 2854, Chemistry by Design**.

If you are taking A2 examinations in **June 2004** you will have been entered for module 2953 **instead** of module 2849. Follow revision for module 2849 but leave out the teaching topic What's in a Medicine? (WM). In addition there will also be a small amount of extra work on alcohols you need to revise. This includes naming, identifying types and the oxidation of alcohols, which can be found in Chemical Ideas 13.2 and 13.4. The effect of complexing on redox reactions (Chemical Ideas 9.3) also needs to be studied for The Steel Story (SS). For details see your specifications. If you are taking module 2854 in June 2004 you will need to revise some extra work not included in this guide. This includes work on soils and the structure of silicates and clays (Chemical Storylines AA2), as well as ion exchange equilibria (Chemical Ideas 7.5), from Aspects of Agriculture (AA).

The table below shows the scheme of assessment for the A2 course (from 2005).

Examination Module	Module title and teaching topics covered	Duration	Number of marks	Mode of assessment	% weighting of Advanced GCE
2849	**Chemistry of Materials** (What's in a Medicine?, Designer Polymers, Engineering Proteins, and The Steel Story)	1hr 30min	90	Written examination	15%
2854	**Chemistry by Design** (Aspects of Agriculture, Colour by Design, The Oceans, Medicines by Design, and Visiting the Chemical Industry)	2 hrs	120	Written examination	20%
2855	**Individual Investigation** (Salters Chemistry)	Not covered in this guide – your centre will assess you	Coursework	15%

The marks gained in the A2 year will be added to those gained in the AS year. The A2 marks count as 50% of your final GCE A level marks.

At the start of each teaching topic in this book there is an introduction which summarises the content. It allows you to see which sections from **Chemical Ideas** and **Chemical Storylines** need to be revised for that specific topic. Some Chemical Ideas in these introductory summaries will be printed in italics. This means you have covered the work elsewhere in the course. It may have been in earlier A2 teaching topics or in the AS course you studied last year.

Within each teaching topic the material is divided into short sections of one to four pages. Each section usually revises one section of material from Chemical Ideas, and ends with **quick check questions** to help you test your understanding.

Also included in the book are sections on **experimental techniques** and **examination hints and tips**. Towards the end of the book there are **practice examination questions**, along with helpful comments to assist in answering.

All questions (quick check and examination questions) are provided with full answers.

What's in a Medicine? (WM)

This unit introduces the importance of the pharmaceutical industry. It uses aspirin as a case study to introduce some important new organic chemistry. The unit also explores how instrumental analysis (mass spectrometry and infrared spectroscopy) can be used to identify a compound. CI refers to sections in your Chemical Ideas textbook.

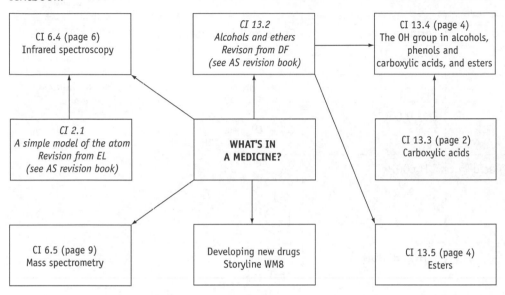

Several parts of this unit do not fit into the main Chemical Ideas sections but are part of the materials you may be examined on. These are:

Development and safety testing of medicines (Storyline WM8)

If the structure of an existing medicine is altered slightly, this will also slightly alter its properties. This means when a useful activity has been identified, chemists will make many similar compounds (called analogues), to try to maximise the desired effects. This development phase may typically take up to 3 years and is very expensive.

The new potential medicine needs to be developed. For example, it needs to be determined how safe the medicine is for human consumption, its stability in the human body and the best form for dispensing, e.g. tablet or aerosol. After extensive testing, which might take up to 6 years, the new medicine is ready for marketing.

As a result of the long time needed to develop and test a new medicine and the fact that, of every 5000 potential new medicines typically only one will ever reach the open market, this becomes a very expensive business costing many millions of pounds. To try to recoup this outlay pharmaceutical companies may do a number of things:

- They will carefully research the demand for their potential new product before spending too much money.

- They will obtain a patent on the new medicine, preventing other companies from producing that medicine for a stated time.

Carboxylic acids
Chemical Ideas 13.3

Carboxylic acids contain the functional group

$$-C\underset{O-H}{\overset{O}{\Large\diagup\!\!\!\!\diagdown}}$$

Their systematic names all end with -oic acid.

Name	Formula	Structural formula
Methanoic acid (one carbon)	HCOOH	$H-C\overset{O}{\underset{OH}{}}$
Propanoic acid	CH_3CH_2COOH	$CH_3-CH_2-C\overset{O}{\underset{OH}{}}$
Benzene carboxylic (benzoic) acid	C_6H_5COOH	
2-methylpropanoic acid	$(CH_3)_2CHCOOH$	
Ethanedioic acid	$(COOH)_2$	

> When counting carbons in the longest chain to name carboxylic acids, remember to include the carbon in the carboxylic acid group.

Reactions

1 Acid–base reactions

Being weak acids, carboxylic acids partially dissociate in aqueous solution to form oxonium ions (H_3O^+) and carboxylate ions ($RCOO^-$):

$$HCOOH(aq) \quad + \quad H_2O(l) \quad \rightleftharpoons \quad H_3O^+(aq) \quad + \quad HCOO^-(aq)$$
methanoic acid   base   oxonium ion   methanoate ion

Carboxylic acids react with bases to produce salts. Chemically they are important because many useful derivatives can be made from them.
(Remember: base + acid → salt + water.)

$$CH_3COOH(aq) \quad + \quad NaOH(aq) \quad \rightarrow \quad CH_3COO^-Na^+(aq) \quad + \quad H_2O(l)$$
ethanoic acid   sodium ethanoate

2 Esterification reactions

Carboxylic acids react with alcohols in the presence of a strong acid catalyst, e.g. a few drops of concentrated sulphuric acid, when heated under reflux. This reaction (called esterification) is reversible and comes to equilibrium during refluxing:

> To simplify things, chemists often say acids produce hydrogen ions (H^+) rather than oxonium ions (H_3O^+).

carboxylic acid   alcohol   ester   water

$$CH_3COOH(l) \;+\; CH_3CH_2OH(l) \;\rightleftharpoons\; CH_3COOCH_2CH_3(l) \;+\; H_2O(l)$$

ethanoic acid   ethanol   ethyl ethanoate

(For further information on esters, see page 5.)

Tests for carboxylic acids

Carboxylic acids are strong enough to react with carbonates to produce carbon dioxide.

(Remember: carbonate + acid → salt + water + carbon dioxide.)

Sodium carbonate or sodium hydrogencarbonate solutions are commonly used to test for acids:

$$2HCOOH(aq) \;+\; Na_2CO_3(aq) \;\rightarrow\; 2HCOO^-Na^+(aq) \;+\; CO_2(g) + H_2O(l)$$

methanoic acid   sodium methanoate

An acid will produce bubbles of carbon dioxide gas, which are readily seen and can be confirmed by testing the gas with limewater, which turns milky:

$$Ca(OH)_2(aq) \;+\; CO_2(g) \;\rightarrow\; CaCO_3(s) \;+\; H_2O(l)$$

cloudy ppt

Derivatives

Carboxylic acids are important chemically because many useful derivatives can be made from them.

Derivative	Functional group	Example	Name
Ester	$\begin{matrix} O \\ \parallel \\ -C-O- \end{matrix}$	$CH_3-\overset{\overset{\textstyle O}{\parallel}}{C}-O-CH_3$	Methyl ethanoate
Acyl chloride	$-C\overset{\textstyle O}{\underset{\textstyle Cl}{<}}$	$CH_3-C\overset{\textstyle O}{\underset{\textstyle Cl}{<}}$	Ethanoyl chloride
Amide	$\begin{matrix} O \\ \parallel \\ -C-N- \\ \mid \\ H \end{matrix}$	$CH_3-\overset{\overset{\textstyle O}{\parallel}}{C}-N\overset{\textstyle H}{\underset{\textstyle H}{<}}$	Ethanamide
Acid anhydride	$\begin{matrix} -C\overset{\textstyle O}{<} \\ \quad\;\; O \\ -C\underset{\textstyle O}{<} \end{matrix}$	$\begin{matrix} CH_3-C\overset{\textstyle O}{<} \\ \qquad\;\; O \\ CH_3-C\underset{\textstyle O}{<} \end{matrix}$	Ethanoic anhydride

> ### ? Quick check questions
>
> 1 Name $CH_3CH_2CH_2CH_2CH_2COOH$.
> 2 Give the equation for propanoic acid reacting with potassium hydroxide.
> 3 Which reactants are needed to make propyl butanoate? (Give names and formulae.)

The OH group in alcohols, phenols and carboxylic acids, and esters

Chemical Ideas 13.4 and 13.5

Phenols

Alcohols contain the hydroxyl functional group (–OH) attached to an aliphatic carbon chain. Some common reactions of alcohols were discussed in the Salters AS revision book.

Phenols are compounds that have one or more –OH groups attached directly to a benzene ring:

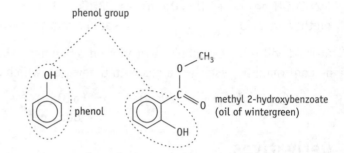

phenol group

phenol

methyl 2-hydroxybenzoate (oil of wintergreen)

Tests for phenols

When **neutral iron(III) chloride** solution is added to phenol or its derivatives a **purple** complex is formed. This test allows us to distinguish between phenols and alcohols.

> Learn the name of the reagent used to identify phenols – this is a common exam question.

Acidic properties of alcohols and phenols

Alcohols, phenols and carboxylic acids are all **weak acids**. They react with water producing oxonium ions. For example:

$$C_6H_5OH \quad + \quad H_2O \quad \rightleftharpoons \quad C_6H_5O^- \quad + \quad H_3O^+$$

phenol   water   phenoxide ion   oxonium ion

The order of acidic strength is:

	ethanol	<	water	<	phenol	<	ethanoic acid
K_a values at 25°C/mol dm^{-3}	1×10^{-16}		1×10^{-14}		1×10^{-10}		1.7×10^{-5}
	weakest acid						strongest acid

> Ethanol is such a weak acid it is even weaker than water! The larger the K_a value, the stronger the acid.

The strength of these compounds as acids can be explained by comparing the **stability of the anion** (R—O$^-$). Phenoxide ions (C$_6$H$_5$O$^-$) and carboxylate ions (RCOO$^-$) are more stable than hydroxide (OH$^-$) and ethoxide ions (CH$_3$CH$_2$O$^-$) because the negative charge on the ion can be **delocalised** across several atoms.

phenoxide ion   carboxylate ion

The difference in acidic strengths of these compounds is illustrated by their reactions with sodium hydroxide and sodium carbonate.

	Ethanol	Phenol	Ethanoic acid
Reaction with NaOH(aq)	No reaction	Reacts to form a salt $NaOH + C_6H_5OH \rightarrow C_6H_5O^-Na^+ + H_2O$	Reacts to form a salt $NaOH + CH_3COOH \rightarrow CH_3COO^-Na^+ + H_2O$
Reaction with Na_2CO_3(aq) (and other carbonates)	No reaction	No reaction	Fizzes – CO_2 released $2CH_3COOH + Na_2CO_3 \rightarrow 2CH_3COO^-Na^+ + CO_2 + H_2O$

Ester formation

Alcohols and phenols can both react to form esters. Alcohols react directly with carboxylic acids in the presence of a concentrated H_2SO_4 catalyst.

ethanol ethanoic acid ethyl ethanoate water

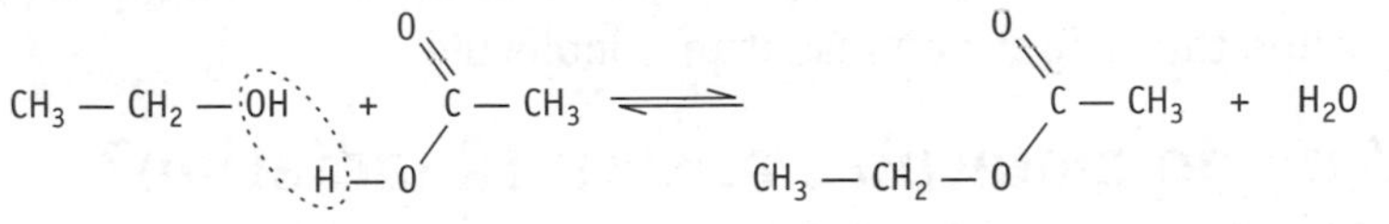

The reaction can be reversed and the ester can be hydrolysed. In this case, a more effective catalyst is dilute sulphuric acid.

Phenols do not form esters in the same way as alcohols. However, phenols will react with **acyl chlorides** to make esters.

phenol ethanoyl chloride phenyl ethanoate hydrogen chloride

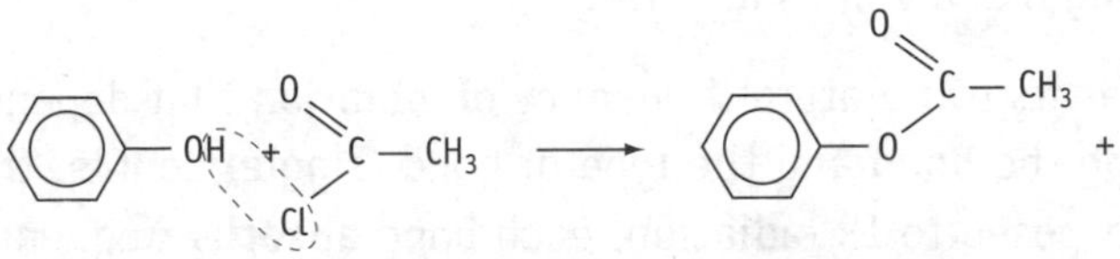

Naming esters

Esters are named in two parts – the first part is from the alcohol (or phenol), and the second part is from the acid (or acid derivative) used to make that ester, e.g. propyl ethanoate:

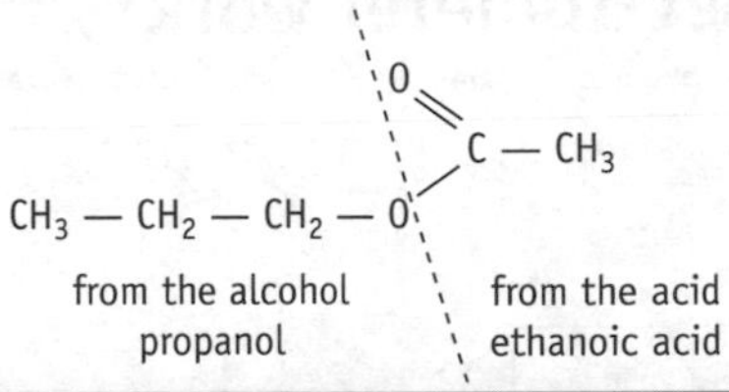

Take care to identify which carbons come from the alcohol and which from the acid. The carbon attached to the carbonyl group is always from the acid.

Esters are also formed when phenols react with acid anhydrides, but you do not need to learn these reactions. You might have used this type of reaction to make aspirin.

? *Quick check question*

1 Study the compounds A–C below.

A

methyl 2-hydroxybenzoate

B

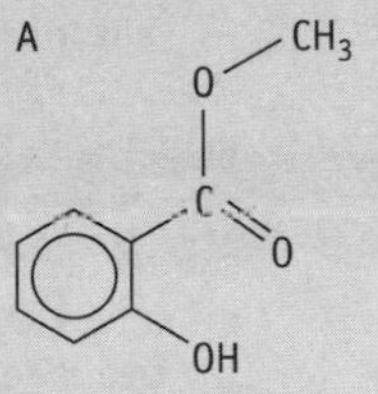

2-hydroxybenzoic acid

C

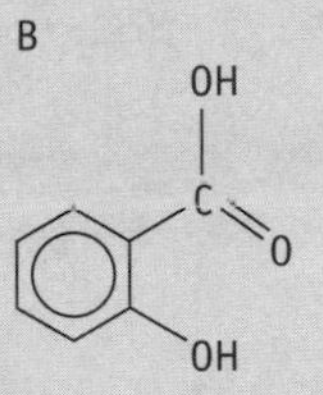

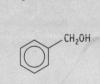

phenylmethanol

(a) Identify the functional groups in each of the molecules.

(b) Place compounds A, B and C in increasing order of acidity.

(c) Suggest how 2-ethanoyloxybenzoic acid (aspirin), whose structural formula is shown below, could be made in a one-step reaction, starting from compound B.

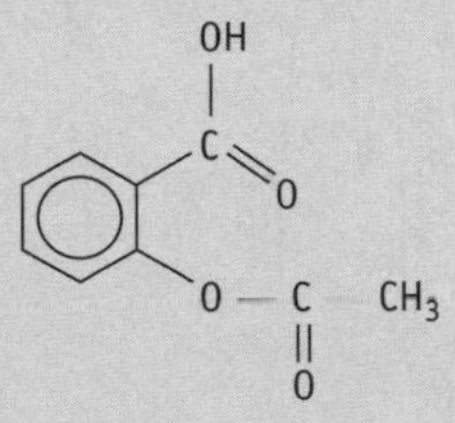

(d) What test could be used to show whether all of compound B had been converted to aspirin?

Infrared spectroscopy
Chemical Ideas 6.4

Infrared (IR) spectroscopy is a useful technique in determining the structure of organic compounds. In particular, it helps chemists to **identify different types of covalent bonds** (e.g. C=O, C=C) and hence draw conclusions about the **functional groups** that might be present in a molecule.

Why do molecules absorb IR radiation?

To understand IR spectroscopy it is helpful to think of the bond between any two atoms as being like a vibrating spring.

Each bond has its own natural frequency of vibration that depends on the types of atoms forming the bond and the type of bond (single, double, triple). When a molecule is exposed to IR radiation, each bond **absorbs** frequencies, causing it to vibrate more vigorously. Different bonds absorb different frequencies of IR radiation.

Vibrating bond in a diatomic molecule

How does an IR spectrometer work?

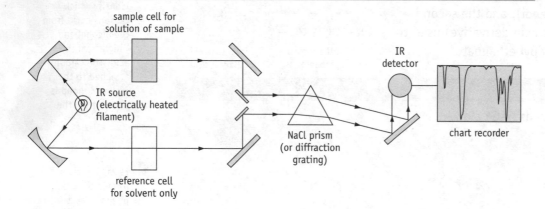

sample cell for solution of sample

IR source (electrically heated filament)

reference cell for solvent only

NaCl prism (or diffraction grating)

IR detector

chart recorder

▶ Identical IR passes through the sample and reference cells. The sample absorbs certain frequencies. Any absorptions by the solvent are detected in the reference cell and subtracted. The difference between the two is recorded on the chart.

A typical IR spectrum

▶ The wavenumber is the number of wavelengths that fit into 1 cm.

The IR spectra below are for ethanol (Spectrum A) and ethanoic acid (Spectrum B), both in the gas phase.

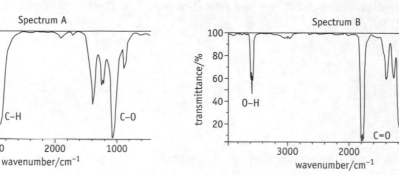

IR spectra have the following features:

- The x-axis shows **wavenumber**, measured in **cm^{-1}**. The scale usually starts at around 4000 wavenumbers on the left and **descends** to about 500 wavenumbers.
- The y-axis shows percentage **transmittance**. Because transmittance is plotted, the baseline is at the top (100% transmittance) and the absorption signals (or bands) are **downward** troughs.
- Absorptions are described as '**strong**', '**medium**' and '**hydrogen bonded**' (**broad**).

Interpreting IR spectra

The most prominent absorption signals (bands) in the IR spectrum can be matched to a particular covalent bond using the table on the right.

For example, in Spectrum B on page 6, the intense absorption in the region of 1700 cm^{-1} is characteristic of the C=O bond.

Bond	Location	Wavenumber/cm^{-1}
C—H	alkanes alkenes, arenes alkynes	2850–2950 3000–3100 ca 3300
C=C	alkenes	1620–1680
⬡	arenes	several peaks in range 1450–1650
C≡C	alkynes	2100–2260
C=O	aldehydes ketones carboxylic acids esters amides	1720–1740 1705–1725 1700–1725 1735–1750 1630–1700
C—O	alcohols, ethers, esters	1050–1300
C≡N	nitriles	2200–2260
O—H	alcohols, phenols *alcohols, phenols *carboxylic acids	3600–3640 3200–3600 2500–3200

> The precise position of an absorption signal depends on the environment of the bond in the molecule, so a wavenumber **region** is quoted in the table.

> You don't need to learn these values. They will be given on the data sheets provided in your exam.

> * The asterisks refer to hydrogen bonded groups.

Alcohol, carboxylic acid, ester or carbonyl compound?

The presence or absence of three key IR absorptions (see table below) should allow you to distinguish between most alcohols, carboxylic acids, esters and carbonyls.

	Is there an absorption in the region 3200–3600 cm^{-1} (caused by hydrogen bonded –OH)?	Is there a strong absorption in the region 1700–1750 cm^{-1} (caused by C=O)?	Is there an absorption in the region 1050–1300 cm^{-1} (caused by C—O)?
Alcohols e.g. CH_3CH_2OH	yes	no	yes
Carboxylic acids e.g. CH_3COOH	yes	yes	yes
Esters e.g. $CH_3COOC_2H_5$	no	yes	yes
Carbonyls e.g. $CH_3CH_2COCH_3$	no	yes	no

Functional groups that are involved in hydrogen bonding (e.g. –OH or –NH in alcohols, phenols, carboxylic acids and amines) usually give broad rather than sharp absorptions.

Worked example

Spectra C and D below are for two compounds, both of which have the formula C_4H_8O. One is butanone and the other is but-2-en-1-ol.

Identify the key absorptions in each spectrum and which bonds they correspond to. Use this information to decide which spectrum represents which compound.

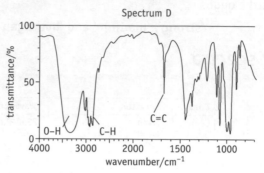

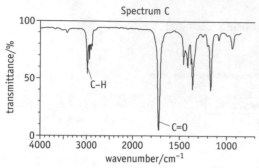

Concentrate on the absorptions above 1500 cm^{-1}.

The characteristic absorptions have been labelled. This allows us to deduce that Spectrum C is for butanone and Spectrum D is for but-2-en-1-ol.

? Quick check questions

1 The IR spectra below are for two compounds A and B that have the same molecular formula, $C_2H_4O_2$. One is a carboxylic acid and the other is an ester.

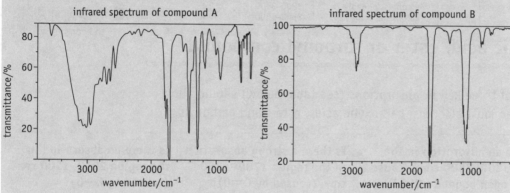

(a) Draw the full structural formula of each compound.

(b) Identify the key absorptions in each spectrum and the bonds to which they correspond. Use this information to match the correct spectrum to each compound.

2 Aspirin can be made in the laboratory by reacting salicylic acid with ethanoic anhydride. How could IR spectroscopy be used to show whether the aspirin produced in a reaction was contaminated with (unreacted) salicylic acid?

Mass spectrometry
Chemical Ideas 6.5

For details on how a mass spectrometer works refer to the Salters AS revision book.

Ionising molecules

Inside the mass spectrometer, bombarding electrons ionise molecules of the sample being tested. For example, if the sample contains butan-2-one molecules:

$$CH_3CH_2COCH_3 + e^- \text{ (high energy)} \rightarrow [CH_3CH_2COCH_3]^+ + 2e^-$$

The ion produced when butan-2-one has lost just one electron is called the **molecular ion (M)**. The molecular ion is unstable and can break up by a process known as **fragmentation** into other fragment ions and uncharged fragments. The formation of the four major fragment ions is shown on the right:

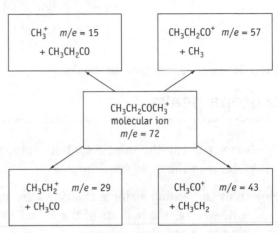

A typical mass spectrum

The mass spectrum of butan-2-one ($CH_3COCH_2CH_3$) is shown below.

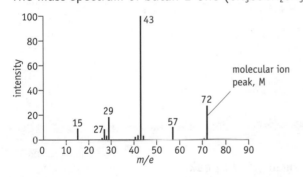

A mass spectrum typically has the following features:

- The x-axis shows **mass-to-charge ratio**, m/e (m for mass, e for charge).
- The y-axis shows **intensity** or percentage abundance.
- Each line on the spectrum represents a **positively charged ion**. One line corresponds to the molecular ion (M); the other lines represent fragments of the molecular ion or isotope peaks (see below).

Interpreting a mass spectrum

Step 1: Identify the molecular ion peak (M). (At low resolution, this is usually the peak with the highest m/e. This peak gives the **relative molecular mass** of the compound.)

▶ m/e stands for 'mass-to-charge ratio'

▶ Only the **ions** formed in fragmentation pass through the mass spectrometer to the detector.

▶ You should assume that all peaks on a mass spectrum are caused by 1^+ ions.

▶ Beware! The molecular ion peak is often very small and the mass spectrum consists mostly of fragments.

Step 2: List the masses of the other major peaks (caused by fragment ions) in the spectrum. Then find the difference in mass between these peaks and the molecular ion peak. Some common peaks to look out for are given in the table on the right.

For example, for butan-2-one:

Molecular ion peak at	A major peak is at...	Difference in mass	Fragment responsible for the major peak
72	57	15 (i.e. loss of CH_3)	$CH_3CH_2CO^+$ or $[C_3H_5O]^+$
72	43	29 (i.e. loss of CH_3CH_2)	CH_3CO^+ or $[C_2H_3O]^+$
72	29	43 (i.e. loss of CH_3CO)	$CH_3CH_2^+$ or $[C_2H_5]^+$
72	15	57 (i.e. loss of CH_3CH_2CO)	CH_3^+

m/e	Possible fragment
15	CH_3^+
17	OH^+
28	$C{=}O^+$ or $C_2H_4^+$
29	$CH_3CH_2^+$
43	CH_3CO^+ or $C_3H_7^+$
77	$C_6H_5^+$

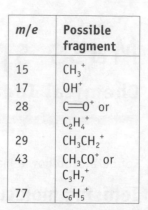
Remember to put the + sign in when writing the formula of a fragment ion.

Step 3: Identify any **isotope peaks**.

Isotope peaks

When molecules in the sample contain isotopes of any atom, additional fragment peaks occur in the mass spectrum.

Molecules containing a single chlorine atom have an isotope peak at M+2, caused by ^{37}Cl. Furthermore, the heights of the M and M+2 peaks are always in a 3 : 1 ratio, mirroring the natural abundances of ^{35}Cl and ^{37}Cl. This is illustrated in the mass spectrum of chloromethane (CH_3Cl), shown below.

Isotope peaks have m/e values larger than that of the molecular ion.

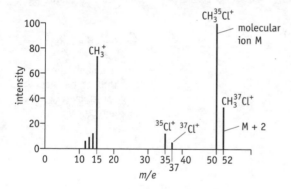

Any fragments containing carbon atoms will always produce a small **(M+1) peak**, caused by the presence of ^{13}C. If a molecule contains only one carbon atom, the ratio of heights of the M and M+1 peaks will be 98.9 : 1.1. For a molecule with two carbon atoms the ratio will be 97.8 : 2.2, and so on.

In a sample of 100 carbon atoms, 98.9% would be ^{12}C and 1.1% would be ^{13}C.

Worked example

The mass spectrum on the right is for an unknown compound P. The relative heights of the peaks at m/e = 120 and 121 are 100 : 8.8. Identify P and suggest the fragments that are responsible for the major peaks identified in the spectrum.

Step 1: The molecular ion peak is probably the peak at m/e = 120. The peak at m/e = 121 is likely to be an isotope peak due to ^{13}C.

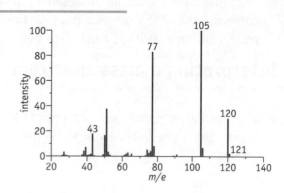

Step 2: There are major peaks at 105, 77 and 43. The peak at 77 is almost certainly due to $C_6H_5^+$. The peak at 43 could be due to either CH_3CO^+ or $C_3H_7^+$. At this stage two molecular formulae fit the data – $C_6H_5COCH_3$ and $C_6H_5C_3H_7$.

Molecular ion peak at	A major peak is at...	Difference in mass	Fragment responsible for the major peak
120	105	15 (i.e. loss of CH_3)	$C_6H_5CH_2CH_2^+$ or $C_6H_5CO^+$
120	77	43 (i.e. loss of CH_3CO^+ or $C_3H_7^+$)	$C_6H_5^+$
120	43	77 (i.e. loss of $C_6H_5^+$)	CH_3CO^+ or $C_3H_7^+$

Step 3: The information about the heights of the M and M+1 peaks suggests that the compound contains eight carbon atoms. This suggests the compound is $C_6H_5COCH_3$.

▶ The absence of peaks at $m/e = 29$ and 91 is further evidence that the compound is not $C_6H_5C_3H_7$.

? Quick check questions

1 The mass spectrum below is that of a common carboxylic acid. Identify the major peaks in the spectrum and find the formula of the acid.

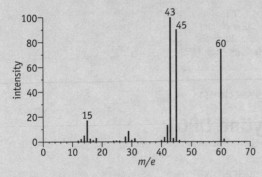

2 The mass spectra of two compounds, A and B, are shown below. Compound A is converted into compound B when refluxed with excess acidified potassium dichromate. No other product is formed. Identify the two compounds and account for the major peaks in each spectrum.

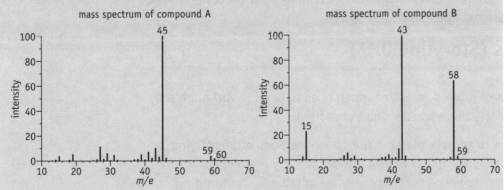

3 Sketch the mass spectrum for the compound bromoethane (C_2H_5Br). Identify the fragment ions responsible for each peak that you draw. (Note: Bromine has two isotopes, ^{79}Br and ^{81}Br, which occur naturally in a 50% : 50% ratio).

Designer Polymers (DP)

This unit studies condensation polymers such as nylons and polyesters and describes how chemists can change polymer properties by modifying their structure. CI refers to sections of your Chemical Ideas textbook.

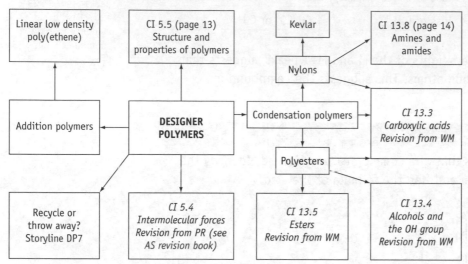

Parts of the Storylines are important right through this unit, especially those highlighted below and Storylines PR2 and PR3.

Some parts of this unit do not fit into the main Chemical Ideas sections:

Historical development of polymers (Storyline DP6)

As the understanding of polymers increased a new era was introduced where chemists were able to specifically design new polymers to meet the demands of everyday life. Wallace Carothers used his understanding of natural proteins such as wool to synthesise synthetic polyamides, called nylons. An understanding of condensation reactions to produce esters led to the development of polyesters, used in a wide range of applications, from clothes to bottles.

Recycling plastics (Storyline DP7)

We generate vast amounts of rubbish, much of which goes to landfill. Although only a small percentage is plastics, these usually don't decompose.

One alternative to burial is to **recycle** plastics, but cheap methods of identifying and sorting different polymers are not currently available. Many thermoplastics are mixtures of monomers – copolymers – that may only be suitable for **remoulding**.

Another approach is to turn polymers back to the original monomers by **cracking**. The recovered monomers can then be turned into new plastics or used in other parts of the chemical industry.

A third option is to design **degradable** plastics. In the future we could use more **biopolymers** (molecules made by living organisms such as bacteria or plants), synthetic **biodegradable** plastics or **photodegradable** plastics. All of these materials break down in the environment, although the speed of this process may still prove to be problematic.

The structure and properties of polymers

Chemical Ideas 5.5 (part 2)

During **condensation polymerisation** two different monomers join together forming a **condensation polymer**. Two common groups of condensation polymers are **polyamides** (nylons) and **polyesters**. Polyamides are made by condensing diamines and dicarboxylic acids. Polyesters are made by condensing diols and dicarboxylic acids.

> ▶ In a condensation reaction, two molecules join together and a smaller molecule, often water, is eliminated.

How do temperature changes affect polymers?

At room temperature, most polymers have some **crystalline** regions (highly ordered chains) and some **amorphous** regions (randomly arranged chains). Cooling the polymer gives more crystalline regions. Eventually, the **glass transition temperature** (T_g) is reached and the polymer becomes brittle or glass-like. As a polymer warms up, it becomes increasingly flexible. After continued heating, the **melting temperature** (T_m) is reached and the polymer becomes liquid.

> ▶ The monomers must have reactive groups at either end of a carbon chain.

> ▶ The higher the percentage of crystalline regions the stiffer the polymer.

Forces between polymer chains

Both nylons and polyesters form linear polymer chains, making them ideal for use as fibres. The individual polymer chains are held together by intermolecular forces.

> ▶ Hydrogen bonding holds the nylon chains together.

> ▶ Permanent dipole interactions hold the polyester chains together.

Modifying polymer properties

Chemists can modify the properties of polymers to suit the needs of the customer, for example by using **copolymerisation** or by adding **plasticisers**.

> ▶ For details of how copolymerisation and plasticisers alter a polymer's properties see page 116 of Chemical Ideas.

? Quick check question

1 Draw a section of the polymer chain formed when the following monomers react:

   (a) $H_2N(CH_2)_4NH_2$ and $HOOC(CH_2)_5COOH$

   (b)

Amines and amides
Chemical Ideas 13.8

Amines are organic derivatives of ammonia. Ammonia has three hydrogen atoms bonded to a central nitrogen atom, whereas amines have one or more alkyl groups substituted for hydrogen. Like ammonia, the **lone pair** of electrons on the nitrogen atom is responsible for the three main properties of amines: they can act as **bases**, **nucleophiles** and **ligands**.

lone pair of electrons

ammonia   methylamine

> The NH_2 functional group is called the **amino** group.

Primary, secondary and tertiary amines

In primary amines, the nitrogen atom is bonded to **one** alkyl (or aryl) group. In secondary amines, the nitrogen atom is bonded to **two** alkyl (or aryl) groups. In tertiary amines, the nitrogen atom is bonded to **three** alkyl (or aryl) groups.

a primary amine   a secondary amine   a tertiary amine

> An aryl group contains a benzene ring with one hydrogen atom substituted, e.g. phenyl, C_6H_5-.

Naming primary amines

Here are the names of some common primary amines:

CH_3NH_2	**methyl**amine
$CH_3CH_2NH_2$	**ethyl**amine
$CH_3CH_2CH_2NH_2$	**propyl**amine
$CH_3CH(NH_2)CH_3$	**2-propyl**amine (in this example, the NH_2 group is attached to the **second** carbon in the propyl chain)
$C_6H_5NH_2$	**phenyl**amine

> For amines with longer alkyl chains, it is equally correct to use the prefix 'amino' followed by the name of the parent alkane, e.g. $CH_3CH_2CH_2CH_2CH_2NH_2$ can be called aminopentane.

Amines are bases

Amines dissolve readily in water forming weakly alkaline solutions:

$$CH_3CH_2CH_2NH_2 \quad + \quad H_2O(l) \quad \rightleftharpoons \quad CH_3CH_2CH_2NH_3{}^+(aq) \quad + \quad OH^-(aq)$$
propylamine   water   propylammonium ion   hydroxide ion

> A **base** is a hydrogen ion (proton) acceptor.

Like ammonia, amines can accept a hydrogen ion from water or from an acid.

$$CH_3CH_2CH_2NH_2 + \quad HCl(aq) \quad \rightarrow \quad CH_3CH_2CH_2NH_3{}^+(aq) \quad + \quad Cl^-(aq)$$
propylamine   hydrochloric acid   propylammonium ion   chloride ion

> The lone pair on the nitrogen atom forms a dative covalent bond with the hydrogen ion.

Amines act as nucleophiles

Primary amines react with halogenoalkanes to form secondary amines, for example:

nucleophilic attack at δ+ carbon

methylamine   chloroethane   N-ethylmethylamine   hydrogen chloride

The primary amine is acting as a nucleophile. It attacks the δ+ carbon in the halogenoalkane. The secondary amine formed in this reaction can go on to react with another molecule of halogenoalkane to make a tertiary amine. Successive hydrogen atoms are being replaced by alkyl groups – this is an **alkylation** reaction.

> ◖ A **nucleophile** is an electron pair donor.

Amines also react with acyl chlorides. The reaction is called an **acylation** reaction and the product is a secondary amide. For example:

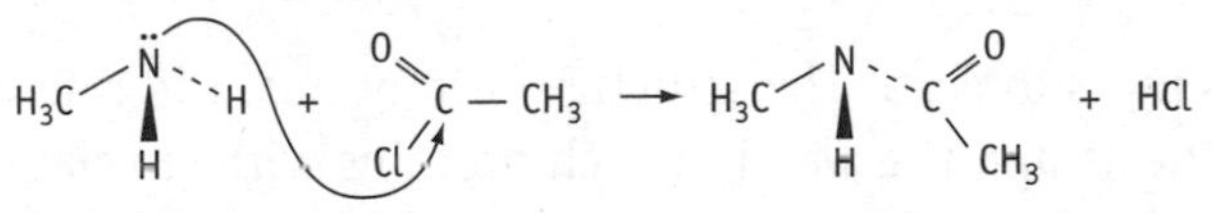

methylamine   ethanoyl chloride   N-methylethanamide   hydrogen chloride

Amides

Primary amides have the structural formula:

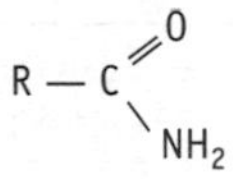

Secondary amides have the structural formula:

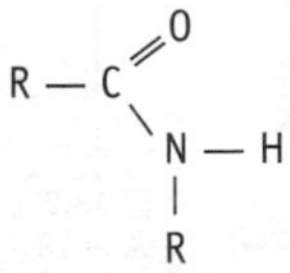

The polymer nylon is an example of a polyamide:

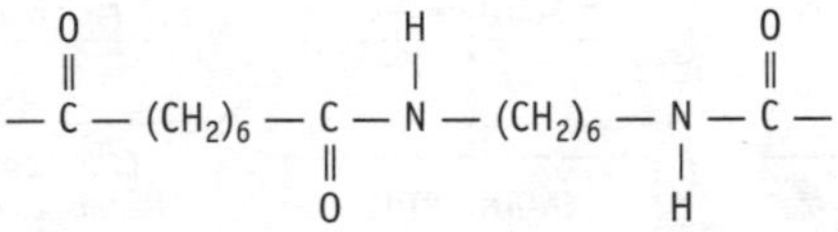

Hydrolysis of amides

Hydrolysis means bond breaking through reaction with water. When amides are hydrolysed, it is the C—N bond that breaks. The reaction can be catalysed by acid or alkali, leading to the formation of slightly different products.

> ◖ Acid hydrolysis gives a carboxylic acid and a substituted ammonium ion while alkaline hydrolysis gives an amine and a carboxylate anion.

An example of acid hydrolysis would be:

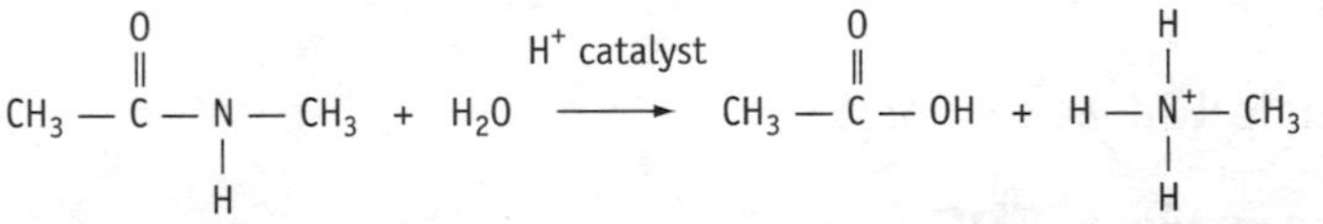

N-methylethanamide   ethanoic acid   ethylammonium ion   moderately concentrated acid (e.g. 4M HCl, reflux)

An example of alkaline hydrolysis would be:

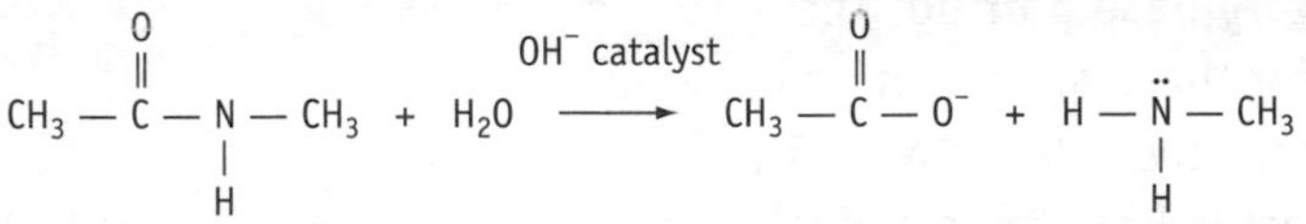

N-methylethanamide   ethanoate ion   ethylamine   moderately concentrated alkali (e.g. 2M NaOH, reflux)

> ◖ Learn the reaction conditions and products for hydrolysis.

Quick check questions

1 Name the following amines:

   (a) CH_3NH_2 (b) $C_6H_5NH_2$ (c) $CH_3CH_2CH_2NH_2$ (d) $CH_3CH_2CH_2(NH)CH_3$

2 Complete the following equations:

   (a) $CH_3CH_2CH_2Cl + CH_3NH_2 \rightarrow$

   (b) $CH_3COCl + CH_3CH_2NH_2 \rightarrow$

3 Write equations for:

   (a) The hydrolysis of $CH_3CONHC_2H_5$ using an acid catalyst.

   (b) The hydrolysis of nylon-6,6 under alkaline conditions.

   For each reaction, give the reagents and essential reaction conditions.

Engineering Proteins (EP)

This units looks at the structure of proteins, the role of RNA and DNA in protein synthesis and the way in which proteins with special properties can affect our lives. CI refers to sections in your Chemical Ideas textbook.

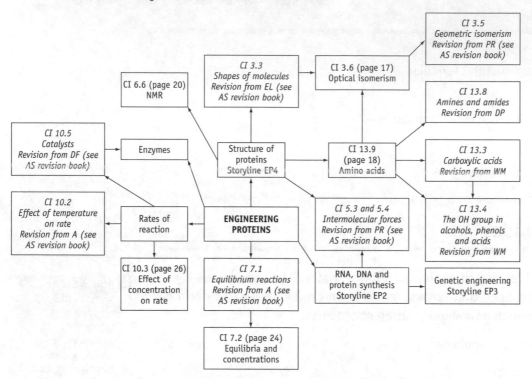

Several parts of this unit do not fit into the main Chemical Ideas sections:

RNA, DNA and protein synthesis (Storyline EP2)

DNA is a double-stranded helix. The strands are held together by hydrogen bonds between pairs of bases. This is known as **complementary base pairing**. The sequence of bases along a strand of DNA is the code for the sequence of amino acids in a protein. After unzipping the DNA the code is transcribed into **mRNA** – this carries a complementary version of the code. mRNA passes from the cell nucleus to the cytoplasm where **ribosomes** translate the code – each three-base **codon** on mRNA codes for one amino acid. Amino acids are brought to the ribosome by specific **tRNA** molecules and are joined together to make the protein chain.

Genetic engineering (Storyline EP3)

To produce human insulin from bacteria:

- A circular piece of bacterial DNA (plasmid) is cut with a restriction enzyme.
- The isolated gene is created and spliced into the plasmid, using other enzymes.
- The modified plasmid is replaced into bacterial cells.
- These modified cells are grown to produce human insulin.
- Remaining cells are destroyed and the protein is harvested.

Uses of genetic engineering include producing a herbicide-resistant maize, plants that produce their own pesticide, oil-eating bacteria and vaccines.

Optical isomerism
Chemical Ideas 3.6

Stereoisomers are molecules that have the same molecular formula and have their atoms bonded in the same order, but these atoms are arranged differently in space. There are two types of stereoisomers – **cis-trans isomers** (for details refer to the Salters AS revision book) and **optical isomers**.

In order to exhibit optical isomerism a molecule must have a **chiral centre** (very often, this central atom is carbon, and is called the **chiral carbon**). Molecules with a chiral centre have non-superimposable mirror images:

The two enantiomers of
2-hydroxypropanoic acid (lactic acid)

> A chiral centre is an atom that has **four different atoms or groups of atoms** attached to it. Molecules with a chiral centre are called **optical isomers** or **enantiomers**.

Identifying the chiral centre in a molecule containing a carbon ring can be tricky. The carbon ring can behave as two different groups if the two 'halves' of the ring are not symmetrical. For example, in carvone:

There are two ways in which optical isomers behave differently from each other (otherwise, optical isomers have identical chemical reactions and physical properties):

- Optically active molecules rotate the plane of plane-polarised light in different directions. One isomer (called the **laevorotatory** or L isomer) rotates light clockwise while the other (the **dextrorotatory** or D isomer) rotates light anticlockwise.

- Optical isomers behave differently in the presence of other chiral molecules. For example, the different smells of oranges and lemons are due to the two optical isomers of the molecule limonene interacting with the chiral receptors in your nose.

> Exam tip: When drawing optical isomers, emphasise the tetrahedral arrangement of groups around the chiral centre by using wedge and dashed bonds.

> To draw enantiomers draw the 3D structure of one enantiomer, then imagine a mirror is placed next to it and draw the reflection.

> To indicate a chiral centre, draw an asterisk next to the relevant carbon atom.

Some chemical reactions produce a 50:50 mixture of D- and L-optical isomers. This type of mixture is called a **racemic mixture** (or DL compound).

? Quick check questions

1 Draw the optical isomers of $CH_3CH(OH)CN$.

2 Identify any chiral centres in the following molecules:

(a)

(b) $CH_3CH(NH_2)COOH$.

Amino acids

Chemical Ideas 13.9 and Storyline EP6

Structure of amino acids

Amino acids are **bifunctional** molecules – they contain both the amino ($-NH_2$) and carboxyl ($-COOH$) functional groups. When these functional groups are attached to the same carbon atom, the amino acid is called an α-amino acid.

All proteins are made from the same 20 α-amino acids. The side chain (R) is different in every amino acid. Each amino acid is also known by a three-letter abbreviation (e.g. Ala for alanine, for which R = CH_3).

Apart from glycine (where R = H), α-amino acids have four different groups attached to the α-carbon atom so they can exhibit **optical isomerism** (see page 17).

▶ You do not need to learn the R groups or abbreviations.

Acid–base reactions

Amino acids can act as both weak acids and weak bases. The $-COOH$ group donates H^+ ions while the $-NH_2$ group accepts H^+ ions. Amino acids can exist in three different ionic forms, depending on the pH of the solution they are in. For example:

▶ The amino group protonates in acidic conditions and the carboxyl group deprotonates in alkaline conditions.

▶ Watch out for amino acids that have $-NH_2$ or $-COOH$ groups as part of the R group. These will also ionise under acidic or alkaline conditions.

Forming dipeptides, polypeptides and proteins

Two amino acids can join together to form a dipeptide. The $-NH_2$ group from one amino acid reacts with the $-COOH$ group from the second amino acid, forming a **secondary amide group** or **peptide link**.

▶ When naming a dipeptide, start at the residue with the free NH_2 group.

▶ Water is eliminated, making this a condensation reaction.

When several amino acids are joined together in this way, a **polypeptide** is formed. **Proteins** are naturally occurring condensation polymers formed when many amino acids join together. The order in which the amino acids join together is called the **primary structure**. When a polypeptide chain forms an alpha helix or beta sheet this is called the **secondary structure**.

▶ Both an alpha helix and a beta sheet are held in place by hydrogen bonding.

The 'global' folding of the polypeptide chain to give it a unique shape is called the **tertiary structure**. The bonding of several polypeptide chains (e.g. in insulin) to form the final protein is called the **quaternary structure**.

The shape is held in place by ionic interactions, covalent bonds and/or hydrogen bonding.

Enzymes (Storyline EP6)

Enzymes are metabolic catalysts that are proteins. They have a high **specificity** for a given substrate. All enzymes have an **active site** where the tertiary structure of the enzyme exactly matches the structure of its substrate (the 'lock and key' mechanism). The substrate can weakly bind to the surface of the active site. This may weaken bonds in the substrate or slightly alter its shape, allowing reaction to occur. After reaction, the products can leave the active site and the process is repeated.

Enzymes are used in many applications, e.g. in diabetes test strips, making cheese and washing powders.

Any changes to the shape of the active site, e.g. disruption of hydrogen bonds on heating or disruption of ionic interactions through changes in pH, will result in the enzyme becoming denatured.

If a molecule of similar shape to the substrate enters the active site it might not react but still occupy that site, preventing entry of the correct substrate. This is called **inhibition**.

Hydrolysis of peptides/proteins

When a peptide or protein is refluxed with moderately concentrated acid or alkali for several hours, the C—N bond in the peptide link is hydrolysed.

Learn the key conditions for hydrolysis, i.e. moderately concentrated (4 mol dm^{-3}) HCl/reflux.

For example:

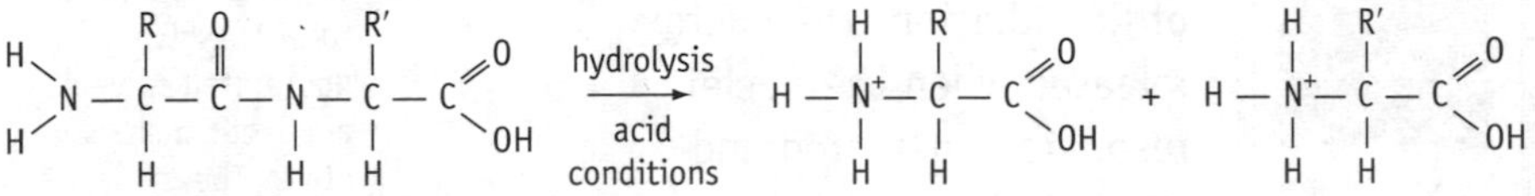

Under acidic hydrolysis conditions the $-NH_2$ groups are protonated to give $-^{+}NH_3$.

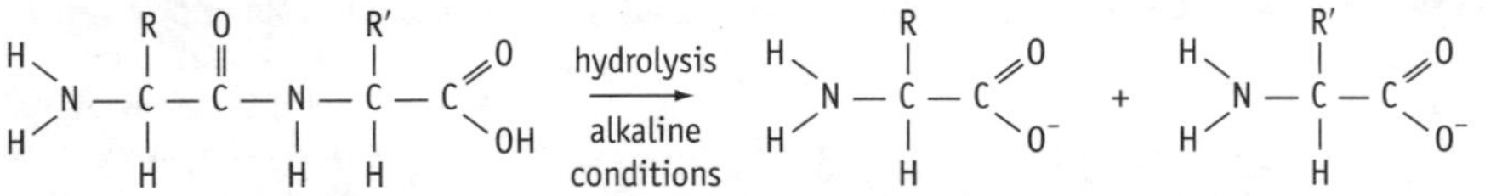

Under alkaline hydrolysis conditions the -COOH deprotonates to give $-COO^{-}$.

? Quick check questions

1. Draw the structures of the products when valine (R = $CH(CH_3)_2$) dissolves in:
   (a) an alkaline solution **(b)** a neutral solution **(c)** an acidic solution
2. Draw the structures of the two different dipeptides that can be produced when glycine (R = H) reacts with serine (R = CH_2OH).
3. Draw the products of the reaction when the tripeptide below is hydrolysed using **(a)** 1 mol dm^{-3} NaOH **(b)** 4 mol dm^{-3} HCl.

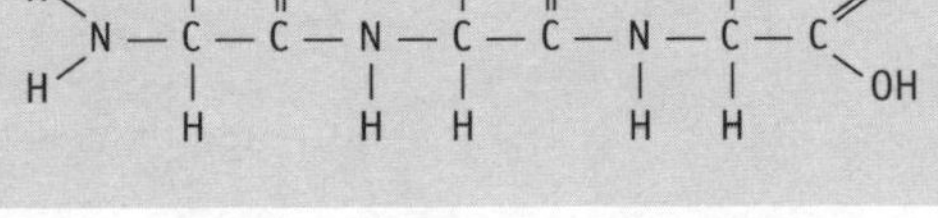

Nuclear magnetic resonance (NMR) spectroscopy

Chemical Ideas 6.6

What is NMR?

The **nuclei** of many atoms have a property called nuclear spin, which makes them behave as if they were tiny bar **magnets**. If these nuclei are placed in a strong magnetic field, some align themselves in the direction of the field (like a compass needle in the Earth's magnetic field) while others align themselves against the field.

Those nuclei aligned in the direction of the field are at a slightly lower energy than those aligned against the field. If the nuclei are given a pulse of radio frequency (RF) radiation, those in the lower energy state are promoted to the higher energy state – this is called **resonance**. The excited nuclei return to their original state by losing the same amount of energy – this is detected by a coil surrounding the sample in the NMR spectrometer.

▶ Some common nuclei that display this property are ^1H, ^{13}C, ^{19}F and ^{31}P.

The NMR spectrometer

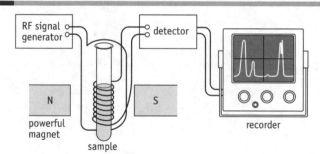

The sample is subjected to pulses of RF radiation. The energy released when the nuclei resonate is detected and converted to an NMR spectrum on the recorder.

▶ If the sample needs to be dissolved, deuterated solvents are used, such as $CDCl_3$. These solvents are used because they do not contain ^1H atoms, and so do not interfere with the spectrum produced.

^{1}H-NMR or proton-NMR

The nucleus most commonly investigated is that of the hydrogen atom ^1H, and when this is the case the technique is known as ^{1}H-NMR or proton-NMR. It provides information about the **chemical environment** in which hydrogen nuclei are found. For example, propanone and propanal both have molecular formula C_3H_6O, but while all six hydrogen nuclei exist in equivalent chemical environments in propanone, there are three different chemical environments for hydrogen in propanal:

The following features are common to all ¹H-NMR spectra:

- **Absorption** is plotted on the y-axis.
- **Chemical shift**, δ, is plotted on the x-axis.
- The chemical shift scale runs from 0 at the right-hand side to around 10.
- There is generally a small absorption caused by a compound called tetramethylsilane (TMS) at the extreme right of the spectrum (δ = 0).

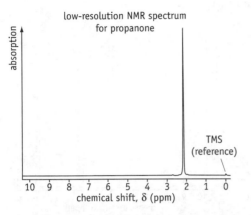

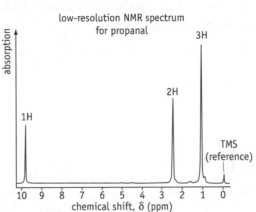

$$CH_3 - Si - CH_3$$

Tetramethylsilane (TMS) gives a strong, sharp absorption peak with a chemical shift well away from other common functional groups. It is used as a reference and assigned a chemical shift of zero.

Interpreting ¹H-NMR spectra

In a low-resolution NMR spectrum there are three things to look for. These are summarised in the following table and explained in detail below.

What to look for	What this tells you
The number of absorption peaks	The number of different chemical environments for ¹H in the molecule.
The position of the absorption peaks (the chemical shift)	The type of protons, e.g. R—CH_3, R—CH_2—R, R—CHO (see below).
The relative area under each absorption peak	The number of equivalent protons in each chemical environment.

Look at the ¹H-NMR spectra for propanone and propanal above.

Number of absorption peaks: The spectrum of propanone has a single absorption peak because all six hydrogen atoms in the molecule are in the same chemical environment. In the spectrum for propanal there are three absorption peaks.

Position of absorption peaks: Read off the chemical shift and then look up this value in the table on page 22. This allows you to identify the type of proton responsible for each absorption peak.

For example, in the spectrum for propanone, the single absorption peak has a chemical shift centred at approximately 2.2. The table of chemical shifts indicates this is caused by R—CO—CH_3.

For propanal, the absorption peak centred at chemical shift 1.1 is caused by R—CH_3, the peak centred at 2.4 is caused by R—CO—CH_2—R and the peak centred at 9.8 is caused by R—CHO.

Relative area under each absorption peak: Modern NMR spectrometers calculate the area under each peak automatically and show this information as an **integration trace**.

You will always be provided with a table of chemical shift values in exams – there is no need to learn this data.

You will not be expected to interpret an integration trace – instead, each absorption peak will be labelled with information about the relative number of protons.

You can see the relative numbers of protons in each chemical environment labelled on the low-resolution NMR spectrum for propanal on page 21.

Worked example

The NMR spectra below are for three isomers with the molecular formula C_3H_8O. Identify the compound responsible for each spectrum.

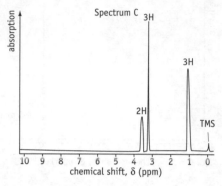

Step 1: Draw all the possible isomers with the formula C_3H_8O. Identify the different chemical environments for hydrogen in each molecule.

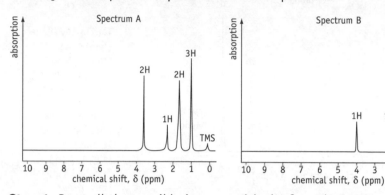

There are three different environments for hydrogen in methoxyethane, four different environments in propan-1-ol and three in propan-2-ol.

Step 2: Count the number of absorption peaks and identify the number of hydrogens responsible for each peak.

Spectrum A has four peaks. It must be the spectrum from propan-1-ol as this is the only isomer of C_3H_8O that has four different environments for hydrogen. The number of hydrogens in each environment is in the ratio 2:1:2:3, which is what we would expect for propan-1-ol.

Spectrum B has three peaks and the ratio of peak areas is 1:1:6. This must be the spectrum for propan-2-ol. It is identified because six hydrogen atoms in this molecule exist in a chemically equivalent environment.

Spectrum C also has three peaks and the ratio of peak areas is 2:3:3. This is the spectrum of methoxyethane.

Type of proton	Chemical shift (δ)
R—CH$_3$	0.8 – 1.2
R—CH$_2$—R	1.4
R—C—H (with R above and below)	1.5
—O—CH$_3$	3.3
—O—CH$_2$—R	3.6
R—CH—OH	4.0
R—CH$_2$—OH	3.6
R—OH	0.5 – 4.5
R—C(=O)—CH$_3$	2.2
R—C(=O)—CH$_2$CH$_3$	2.4
R—C=O with H	10.0
R—C(=O)—OH	9 – 15
R—C(=O)—O—CH$_3$	3.7

Approximate chemical shift values for some types of proton

Step 3: Explain the position of each peak using chemical shift data. The table below shows how each peak has been matched to a chemical shift value.

Spectrum A		Spectrum B		Spectrum C	
Propan-1-ol		Propan-2-ol		Methoxyethane	
Type of proton	Chemical shift	Type of proton	Chemical shift	Type of proton	Chemical shift
R—CH$_2$—O	3.6	R—CH—OH	4.0	R—O—CH$_3$	3.3
R—OH	2.3	R—OH	2.1	R—CH$_2$—O	3.6
R—CH$_2$—R	1.6	R—CH$_3$	1.2	R—CH$_3$	1.0
R—CH$_3$	0.9				

? Quick check questions

1 How many different absorption peaks would you expect to find in the low-resolution ^{1}H-NMR spectrum of

   (a) methanol **(b)** benzene **(c)** propane **(d)** ethanoic acid?

   What would the relative areas under each peak be for these four compounds?

2 **(a)** Draw the full structural formula of ethanol and identify the different types of proton in the molecule.

   (b) Sketch the NMR spectrum you would expect for ethanol. Use the table of chemical shifts on page 22 to ensure that the absorption peaks you draw are centred on the correct chemical shift value.

3 The NMR spectrum below is of a compound with molecular formula $C_4H_8O_2$. Suggest, with reasons, what the structural formula of the compound is.

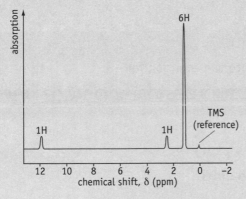

Equilibria and concentrations
Chemical Ideas 7.2

Consider the general equilibrium reaction:

$$aA(aq) + bB(aq) \rightleftharpoons cC(aq) + dD(aq)$$

The **equilibrium law** states that $\quad K_c = \dfrac{[C]^c[D]^d}{[A]^a[B]^b}$

Remember – products divided by reactants!

This constant K_c is the **equilibrium constant** for the reaction. The letter K is used to represent *all* equilibrium constants. When the expression is written in terms of concentrations, we write K_c.

For example, for the equilibrium reaction:

$$3H_2(g) + N_2(g) \rightleftharpoons 2NH_3(g)$$

$$K_c = \frac{[NH_3]^2}{[H_2]^3[N_2]}$$

K_c is a measure of **how far** a reaction proceeds. If the equilibrium mixture is largely composed of reactants, the value of K_c is small, whereas if the equilibrium mixture is largely composed of products, the value of K_c is large.

> ● a and b are the number of moles of reactants A and B; c and d are the number of moles of products C and D.
>
> ● You will probably come across lots of other equilibrium constants – K_p, K_a, K_{sp}.
>
> ● Square brackets around a formula means the concentration of whatever is inside the brackets.

Units of K_c

Example 1:

$$CH_3COOH(l) + C_2H_5OH(l) \rightleftharpoons CH_3COOC_2H_5(l) + H_2O(l)$$

The units of concentration are mol dm^{-3}. Substituting mol dm^{-3} into the equation for K_c allows you to find the units of K_c:

$$K_c = \frac{[CH_3COOC_2H_5(l)][H_2O(l)]}{[CH_3COOH(l)][C_2H_5OH(l)]}$$

$$\text{units} = \frac{(\text{mol dm}^{-3})(\text{mol dm}^{-3})}{(\text{mol dm}^{-3})(\text{mol dm}^{-3})} \quad \text{this cancels out to } \textbf{no units}$$

Example 2:

If $K_c = \dfrac{[NH_3]^2}{[H_2]^3[N_2]}$

$$\text{units} = \frac{(\text{mol dm}^{-3})^2}{(\text{mol dm}^{-3})^3(\text{mol dm}^{-3})} = \frac{1}{(\text{mol dm}^{-3})^2} = \text{mol}^{-2}\text{dm}^6$$

Calculations involving K_c

Example 1:

Calculating the value of K_c when given information on equilibrium concentrations.

Calculate the value of K_c at 763 K for the reaction $H_2(g) + I_2(g) \rightleftharpoons 2HI(g)$ given the following data:

$[H_2(g)] = 1.92$ mol dm^{-3}, $[I_2(g)] = 3.63$ mol dm^{-3} and $[HI(g)] = 17.8$ mol dm^{-3}

$$K_c = \frac{[HI]^2}{[H_2][I_2]} = \frac{17.8^2}{1.92 \times 3.63} = 45.5 \text{ (no units)}$$

Example 2:

Calculating the composition of equilibrium mixtures:

$$CH_3COOH(l) + C_2H_5OH(l) \rightleftharpoons CH_3COOC_2H_5(l) + H_2O(l)$$

K_c for the above esterification reaction has a value of 4.1 at 25°C. Calculate the equilibrium concentration of ethyl ethanoate, given that $[CH_3COOH(l)] = 0.255$ mol dm^{-3}, $[C_2H_5OH(l)] = 0.245$ mol dm^{-3} and $[H_2O(l)] = 0.437$ mol dm^{-3}.

$$K_c = \frac{[CH_3COOC_2H_5(l)][H_2O(l)]}{[CH_3COOH(l)][C_2H_5OH(l)]}$$

> Remember – if data is given to 3 significant figures, you should give your answer to 3 significant figures.

Substitute the values into the expression for K_c:

$$4.1 = \frac{[CH_3COOC_2H_5(l)] \times (0.437)}{(0.255) \times (0.245)}$$

Rearranging the expression:

$$[CH_3COOC_2H_5(l)] = \frac{4.1 \times (0.255) \times (0.245)}{(0.437)} = 0.586 \text{ mol dm}^{-3}$$

What affects the value of K_c?

The only thing that affects the value of K_c is a change in **temperature.**

	Exothermic reactions	**Endothermic reactions**
Temperature increases	Value of K_c decreases	Value of K_c increases
Temperature decreases	Value of K_c increases	Value of K_c decreases

? *Quick check questions*

1 Write expressions for K_c for the following equilibrium reactions. Include the units of K_c in each case:

(a) $2SO_2(g) + O_2(g) \rightleftharpoons 2SO_3(g)$

(b) $N_2O_4(g) \rightleftharpoons 2NO_2(g)$

2 Calculate the value of K_c for the Haber process reaction, $3H_2(g) + N_2(g) \rightleftharpoons 2NH_3(g)$ at 1000 K, given $[H_2(g)] = 1.84$ mol dm^{-3}, $[N_2(g)] = 1.36$ mol dm^{-3} and $[NH_3(g)] = 0.142$ mol dm^{-3}.

3 Ethanol is produced in industry by the hydration of ethene: $C_2H_4(g) + H_2O(g) \rightleftharpoons C_2H_5OH(g)$. The forward reaction is exothermic.

(a) Explain how increasing the pressure would affect the position of equilibrium and the value of K_c.

(b) Explain how increasing the temperature would affect the position of equilibrium and the value of K_c.

The effect of concentration on rate

Chemical Ideas 10.3

What do we mean by rate of reaction?

The rate of reaction is a measure of how fast reactants are used up or how fast products are formed.

$$\text{Rate of reaction} = \frac{\text{change in concentration of reactants or products}}{\text{time}}$$

> ◖ The units of rate of reaction are usually **mol dm^{-3} s^{-1}**.

What is a rate equation?

For the general reaction: A + B → products

The rate equation would be: **rate = k [A]^m [B]^n**

where:

- [A] and [B] are the initial concentrations of reactants A and B
- k is the rate constant for the reaction
- m is the order of reaction with respect to reactant A
- n is the order of reaction with respect to reactant B
- ($n + m$) is the overall order of the reaction

> ◖ You will only encounter values of m and n that are 0, 1 or 2.

> ◖
> | Order with respect to reactant | Effect on rate of reaction |
> | --- | --- |
> | zero | nil |
> | first | doubles |
> | second | quadruples |

Some examples of rate equations are given below:

CH$_3$COCH$_3$(aq) + I$_2$(aq) $\xrightarrow{\text{H}^+}$ CH$_3$COCH$_2$I(aq) + H$^+$(aq) + I$^-$(aq) **rate = k[CH$_3$COCH$_3$][H$^+$]** The reaction is: first order with respect to CH$_3$COCH$_3$ first order with respect to H$^+$ zero order with respect to I$_2$ second order overall	(CH$_3$)$_3$CBr(aq) + OH$^-$(aq) → (CH$_3$)$_3$COH(aq) + Br$^-$(aq) **rate = k[(CH$_3$)$_3$CBr]** The reaction is: first order with respect to (CH$_3$)$_3$CBr zero order with respect to OH$^-$ first order overall

The effect of a temperature change on the value of k

The rate of a chemical reaction increases whenever temperature is increased:

$$\text{rate} = k[A]^m[B]^n$$

Therefore, a rise in temperature increases the value of the rate constant, k.

> ◖ Remember: it is the initial concentrations of the reactants (not products!) that appear in the rate equation.

> ◖ Catalysts are not classed as reactants in chemical equations but they **can** appear in a rate equation.

Determining a rate equation

A rate equation needs to be determined experimentally. If more than one reactant is involved, a series of experiments needs to be carried out, e.g. for the following reaction:

$$A + B \rightarrow products$$

- An initial set of reactions is carried out varying only the concentration of reactant A, keeping the concentration of reactant B the same each time.
- The process is repeated, this time varying the concentration of reactant B and keeping the concentration of reactant A the same each time.
- For each reaction the rate needs to be determined. This can be done by measuring a suitable property, e.g. change in colour using a colorimeter, pH change or volume of gas evolved.

There are various ways to process the experimental data, once it has been collected.

1 Using a concentration–time graph (a **progress curve**)

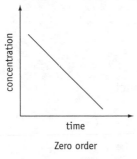

Zero order

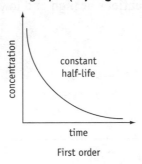

First order

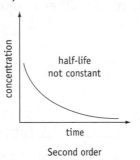

Second order

2 Using the half-lives method

In order to confirm that the reaction is first order, the **half-life** ($t_{1/2}$) can be found.

A number of half-lives need to be determined (see graph on the right) and compared. If the half-life is constant the reaction is first order with respect to that reagent.

3 Using rate–concentration graphs

A **rate–concentration graph** can be constructed by plotting the rate of a reaction against the concentration of a reactant (often referred to as a rate–concentration graph) (these results can be obtained from a series of experiments or by determining the different rates as a reaction progresses). The shape of the graph tells you the order with respect to a reactant.

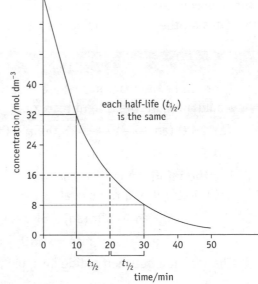

each half-life ($t_{1/2}$) is the same

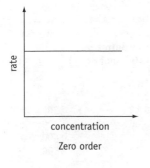

Zero order

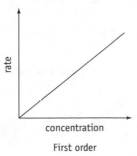

First order

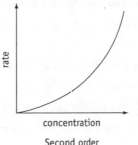

Second order

> ▶ The process is repeated for every reactant involved.

> ▶ It is also important to keep the temperature constant during the experiment, or k will change.

> ▶ The half-life ($t_{1/2}$) is the time taken for the concentration of the reactant to halve.

> ▶ A plot of rate vs (concentration)^2 that gives a straight line indicates a second-order reaction.

Reaction mechanisms

A reaction mechanism describes the step-by-step way in which a chemical reaction occurs.

The slowest step in a multi-step reaction is called the **rate-determining step**. The rate equation for a reaction tells us which particles are involved in the rate-determining step, e.g. consider the following reaction:

$$(CH_3)_3CBr(aq) + OH^-(aq) \rightarrow (CH_3)_3COH(aq) + Br^-(aq)$$

$$rate = k[(CH_3)_3CBr]$$

The reaction is first order with respect to 2-bromo-2-methylpropane and zero order with respect to hydroxide ion. This means that the rate-determining step involves only 2-bromo-2-methylpropane.

> ▶ An intermediate is a chemical formed and then destroyed during the course of the reaction.

> ▶ By studying rate equations and orders, chemists can deduce a mechanism for a reaction.

Reactions involving enzymes

Unusually, enzyme-catalysed reactions have different rate equations at high and low substrate concentrations.

For the reaction:

E	+	S	⟶	ES	⟶	EP	⟶	E	+	P
enzyme		substrate		enzyme substrate complex		enzyme product complex		enzyme		product

At low concentrations of substrate the rate equation is: $rate = k[E][S]$

At high concentrations of substrate the rate equation is: $rate = k[E]$, because all the active sites on the enzyme molecules have become **saturated**.

? *Quick check questions*

1 For each of the reactions whose equation is given below, suggest how you would measure the rate of reaction:

   (a) $2H_2O_2(aq) \xrightarrow{\text{catalase}} 2H_2O(aq) + O_2(g)$

   (b) $BrO_3^-(aq) + 5Br^-(aq) + 6H^+(aq) \rightarrow 3Br_2(aq) + 3H_2O(aq)$

2 Use the information given below to write down the order with respect to each reactant and the overall order for the reaction:

   $$BrO_3^-(aq) + 5Br^-(aq) + 6H^+(aq) \rightarrow 3Br_2(aq) + 3H_2O(aq)$$

   $$rate = k[BrO_3^-][Br^-][H^+]^2$$

3 The reaction between iodide ions and peroxodisulphate ions:

   $$S_2O_8^{2-}(aq) + 2I^-(aq) \rightarrow 2SO_4^{2-}(aq) + I_2(aq)$$

   is first order with respect to each reactant. Write the rate equation for the reaction.

4 The table to the right shows data from an experiment to investigate the rate at which an ester is hydrolysed by alkali.

   (a) Plot a graph of concentration of ester against time.

   (b) Determine two values for the half-life of the reaction.

   (c) What is the order of this reaction with respect to ester?

Ester concentration (10^{-2} mol dm^{-3})	Time(s)
1.00	0
0.68	100
0.48	200
0.30	300
0.22	400
0.12	500

The Steel Story (SS)

Steel is one of the world's most versatile metals. This unit looks at how steel is made and why it rusts. It also takes a detailed look at the properties of iron and other transition metals. CI refers to sections in your Chemical Ideas textbook.

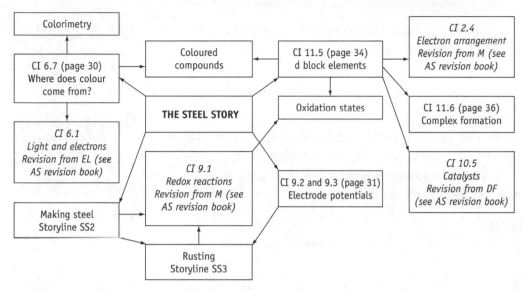

Several parts of this unit do not fit into the main Chemical Ideas sections:

Basic Oxygen Steelmaking (BOS) process (Storyline SS2)

The percentage of elements such as C, Si, Mn, P and S must be carefully controlled to produce steel of the correct specification. In a series of redox reactions sulphur is removed first, by blowing magnesium into the molten iron. The carbon content of iron from the blast furnace is lowered by blowing in high-pressure oxygen; the oxygen also reacts with Si and P forming oxides. These acidic oxides are reacted with a basic oxide (CaO) to form a slag. Finally, other metals can be added if alloy steels are required.

$$Mg + S \rightarrow MgS$$
$$2C + O_2 \rightarrow 2CO$$
$$Si + O_2 \rightarrow SiO_2$$
$$4P + 5O_2 \rightarrow P_4O_{10}$$
$$CaO + SiO_2 \rightarrow CaSiO_3$$

The final composition of the steel is critical for it to perform the required tasks, e.g. high levels of chromium are classed as 'stainless steel' and makes the steel suitable for use as cutlery.

Why recycle steel? (Storyline SS4)

About 40% of the world's steel is recycled. This saves energy, mineral resources and helps reduce pressure on landfill sites. Scrap steel is also important in the BOS process – it is added to the converter *before* the molten iron is poured in to help reduce 'thermal shock'.

Manganate(VII) titrations (Activity SS1.2)

Manganate(VII) ions can be used to analyse titrimetrically for oxidisable material such as Fe^{2+} ions in solution. The reaction is self-indicating (titrate to the first signs of a persistent pink colour).

Where does colour come from?

Chemical Ideas 6.7

Why does an object appear coloured?

If visible light falls on a coloured object, some wavelengths are absorbed and some are reflected or transmitted. What we see are the reflected or transmitted wavelengths.

For example, copper sulphate solution appears blue because it transmits blue light and absorbs all colours other than blue.

A **colour wheel** can help us to work out which colour(s) are most easily absorbed by an object. A blue object absorbs most strongly in the orange region of the visible spectrum. Orange is called the **complementary colour** to blue and is opposite blue in the colour wheel. It is the complementary colour that is seen.

Electronic transitions

When visible light falls on a coloured molecule, the absorbed light is in the energy range that causes **electronic transitions** – electrons move to higher energy levels and the molecule becomes **'excited'**. Molecules do not remain excited for long: as electrons fall back to intermediate energy levels energy is re-emitted in various forms including vibrational energy.

increasing energy

excited electronic level

different vibrational energy levels

ground electronic level

X_2*

X_2

increasing vibrational energy in an excited electronic level

increasing vibrational energy in ground electronic level

Colorimetry

Colorimetry is an experimental technique used to find the concentration of a coloured solution. The amount of light absorbed by a solution – its **absorbance** – is proportional to the concentration of the solution. In a colorimeter, a narrow beam of light passes through a filter towards a tube containing the coloured solution. Any light that passes through the solution is detected by a photocell and a reading of absorbance is displayed on the meter. For details of how to use a colorimeter see page 76.

> Visible light has wavelengths between 400 nm (violet) and 700 nm (red).

> If **all** the visible light falling on a solution is transmitted, it will appear colourless.

> It is useful to learn the colour wheel.

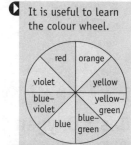

> Ultraviolet light also causes electron transitions in molecules (see page 46).

> Exam tip – learn the essential steps in carrying out a colorimetry experiment.

? Quick check questions

1 What colour light would be most strongly absorbed by:

   (a) a solution of Co^{2+}(aq), which is pink?

   (b) a solution of Fe^{3+}(aq), which is yellow?

2 When setting up a colorimeter, why should you select a filter with the complementary colour to the solution being tested?

Electrode potentials

Chemical Ideas 9.2 and 9.3 and Storyline SS3

Redox reactions

Redox reactions can be considered as two different reactions occurring simultaneously. One is a reduction and the other an oxidation. The equation for each is called a half-equation.

Combining half-equations

Adding together the two half-equations gives the overall (ionic) equation for any redox reaction.

Step 1: Write the half-equations for the oxidation and reduction reactions:

$$Fe^{2+}(aq) \rightarrow Fe^{3+}(aq) + e^- \qquad \textbf{oxidation half-reaction}$$
$$MnO_4^-(aq) + 8H^+(aq) + 5e^- \rightarrow Mn^{2+}(aq) + 4H_2O(l) \quad \textbf{reduction half-reaction}$$

Step 2: Make the number of electrons the same in each half-equation:

$$\mathbf{5Fe^{2+}(aq) \rightarrow 5Fe^{3+}(aq) + 5e^-}$$
$$MnO_4^-(aq) + 8H^+(aq) + 5e^- \rightarrow Mn^{2+}(aq) + 4H_2O(l)$$

Step 3: Add the two half-equations together. The electrons don't appear in the final ionic equation, since they cancel out:

$$5Fe^{2+}(aq) + MnO_4^-(aq) + 8H^+(aq) \rightarrow 5Fe^{3+}(aq) + Mn^{2+}(aq) + 4H_2O(l)$$

Electrode potentials

When a metal is placed in an aqueous solution of its ions, an **equilibrium** is established. A potential difference or **electrode potential** is created between the metal and the solution of ions, e.g.

$$Cu^{2+}(aq) + 2e^- \rightleftharpoons Cu(s)$$
$$Fe^{2+}(aq) + 2e^- \rightleftharpoons Fe(s)$$

The greater the tendency of the metal to release electrons and form ions, the more negative the electrode potential. Altering the temperature or the concentration of ions in solution affects the electrode potential. When finding the potential difference between two half-cells, measurements must be made under **standard conditions**.

Electrochemical cells

It is not possible to measure the electrode potential of a single half-cell. Instead, two half-cells are connected together to make an **electrochemical cell**. The potential difference between the two half-cells is called the cell potential or emf, E_{cell}.

Redox reactions involve transfer of electrons. Remember OILRIG: **O**xidation **is l**oss of electrons. **R**eduction **is g**ain of electrons. A reducing agent **donates electrons** while an oxidising agent **accepts electrons**.

Electrode potentials (see later) allow you to decide which is the oxidation half-reaction and which is the reduction half-reaction.

The Fe^{2+}/Fe redox couple has a more negative electrode potential than the Cu^{2+}/Cu redox couple. Notice the oxidised species is always written first.

Standard conditions are:
Temperature 298 K
Pressure 1 atm
Concentration
1.00 mol dm^3
(all ions).

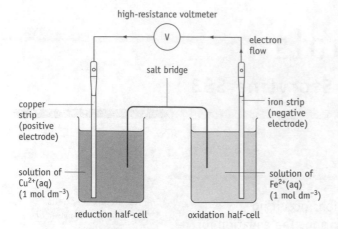

When two half-cells are joined:

- The one with the more negative electrode potential becomes the negative terminal of the electrochemical cell.
- The one with the more positive electrode potential becomes the positive terminal.
- Electrons flow in the external circuit from the negative terminal to the positive terminal.

> The **salt bridge** provides an ionic connection between half-cells. It is usually made from a strip of filter paper soaked in a saturated solution of potassium nitrate.

> A high-resistance voltmeter measures the maximum potential difference between the two half-cells – the cell potential or cell emf, E_{cell}.

Standard electrode potentials

The **standard hydrogen half-cell** (shown on the right) is chosen as the reference electrode against which all other electrode potentials are measured. Its electrode potential under standard conditions is defined as 0.00 volts.

The half-reaction occurring in the standard hydrogen half-cell is:

$$H^+(aq) + e^- \rightleftharpoons \tfrac{1}{2}H_2(g)$$

The **standard electrode potential**, $E^{\ominus}$, of any half-cell is defined as the potential difference between it and the standard hydrogen half-cell.

By convention, the half-reactions are always written as reduction processes (i.e. with the oxidised species and electrons on the left-hand side).

To measure a standard electrode potential, the half-cell being investigated is connected to the standard hydrogen half-cell. For half-cells involving molecules and ions ($I_2/2I^-$) or ions (Fe^{3+}/Fe^{2+}), an inert electrode such as platinum is dipped into a solution containing all the ions and molecules involved in the half-reaction. An Fe^{3+}/Fe^{2+} half-cell is shown on the right.

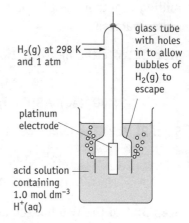

A standard hydrogen half-cell

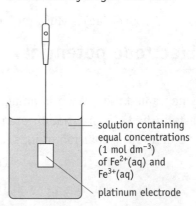

A half-cell involving two different ions

Finding E_{cell}

For example, to find E_{cell} when the Fe^{2+}/Fe and Cu^{2+}/Cu half-cells are connected.

Step 1: Look up the standard electrode potentials for the two half-reactions.

Step 2: Construct an electrode potential chart (see right). The half-cell with the most positive electrode potential is at the bottom of the chart.

Step 3: Find the difference between the two standard electrode potentials. This is E_{cell}.

Alternatively, $E_{cell} = E$ (positive electrode) – E (negative electrode).

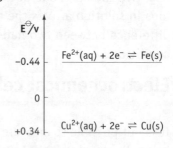

> The E_{cell} value is always positive.

Predicting the direction of a reaction

The key idea to remember is that the half-cell with the more negative electrode potential supplies electrons to the half-cell with the more positive electrode potential.

For example, to write an equation for the reaction, if any, that will occur when aqueous chlorine is added to potassium iodide solution and calculate E_{cell}:

Step 1: Look up the half-reactions and their standard electrode potentials:

$$I_2(aq) + 2e^- \rightleftharpoons 2I^-(aq) \quad E^\ominus = +0.54 \text{ V} \quad \text{half-reaction 1}$$

$$Cl_2(g) + 2e^- \rightleftharpoons 2Cl^-(aq) \quad E^\ominus = +1.36 \text{ V} \quad \text{half-reaction 2}$$

Step 2: Identify which half-reaction has the more negative electrode potential. Rewrite the half-reaction to show it **supplying** electrons (i.e. with the electrons on the right-hand side).

Half-reaction 1 has the more negative $E^\ominus$. This half-reaction is rewritten:

$$2I^-(aq) \rightarrow I_2(aq) + 2e^-$$

Step 3: Balance the number of electrons and then add half-reactions:

$$2I^-(aq) + Cl_2(g) \rightarrow I_2(aq) + 2Cl^-(aq) \quad \text{(in this case electrons already balance)}$$

$E_{cell} = E \text{ (positive electrode)} - E \text{ (negative electrode)} = +1.36 - (+0.54) = +0.82 \text{ V}$

> This method can be used to predict the feasibility of a reaction occurring. The reaction may not occur because the activation enthalpy is high. A catalyst might allow the reaction to occur. Alternatively, this method gives us the feasibility of a reaction under standard conditions. Changing factors such as concentration and/or temperature may cause the reaction to happen.

Corrosion prevention

Rusting is a redox reaction. The two half-reactions involved in the initial stage are:

$$Fe^{2+}(aq) + 2e^- \rightleftharpoons Fe(s) \qquad E^\ominus = -0.44 \text{ V}$$

$$O_2(g) + 2H_2O(l) + 4e^- \rightleftharpoons 4OH^-(aq) \qquad E^\ominus = +0.40 \text{ V}$$

One method of preventing rusting is to attach a piece of zinc to the iron object:

$$Zn^{2+}(aq) + 2e^- \rightleftharpoons Zn(s) \quad E^\ominus = -0.76 \text{ V}$$

> The Zn^{2+}/Zn half-reaction has a more negative electrode potential than the Fe^{2+}/Fe half-reaction and therefore releases electrons more readily. The zinc is oxidised instead of the iron.

? Quick check questions

1 Combine the following half-equations to produce a balanced ionic equation:

   (a) $MnO_4^-(aq) + 8H^+(aq) + 5e^- \rightarrow Mn^{2+}(aq) + 4H_2O(l)$ and
   $2Cl^-(aq) \rightarrow Cl_2(g) + 2e^-$

   (b) $H_2O_2(aq) + 2H^+ + 2e^- \rightarrow 2H_2O(l)$ and $2I^-(aq) \rightarrow I_2(aq) + 2e^-$

2 Describe how you would measure experimentally the standard electrode potential for:

   (a) the Cu^{2+}/Cu half-cell **(b)** the $Cl_2/2Cl^-$ half-cell.

3 Calculate E_{cell} for the electrochemical cells made by combining the following half-cells:

   (a) Zn^{2+}/Zn and Fe^{3+}/Fe^{2+}

   (b) Cu^{2+}/Cu and Ag^+/Ag

4 Use the data below to explain whether a reaction will occur between the following pairs of chemicals:

   (a) Fe^{2+} and Br^- **(b)** Fe^{2+} and $Cr_2O_7^{2-}$ **(c)** Br^- and $Cr_2O_7^{2-}$

   $Cr_2O_7^{2-}(aq) + 14H^+(aq) + 6e^- \rightleftharpoons 2Cr^{3+}(aq) + 7H_2O(l) \qquad E^\ominus = +1.36 \text{ V}$

   $Br_2(aq) + 2e^- \rightleftharpoons 2Br^-(aq) \qquad E^\ominus = +1.07 \text{ V}$

   $Fe^{3+}(aq) + e^- \rightleftharpoons Fe^{2+}(aq) \qquad E^\ominus = +0.77 \text{ V}$

Half-cell	$E^\ominus$/V
$Zn^{2+}(aq)/Zn(s)$	−0.76
$Fe^{3+}(aq)/Fe^{2+}(aq)$	+0.77
$Cu^{2+}(aq)/Cu(s)$	+0.34
$Ag^+(aq)/Ag(s)$	+0.80

The d block: transition metals
Chemical Ideas 11.5

The transition metals are found in the d block of the Periodic Table. The d block contains 30 elements, arranged in three rows of ten. This chapter focuses on the properties of the first row of ten elements, from scandium (Sc) to zinc (Zn).

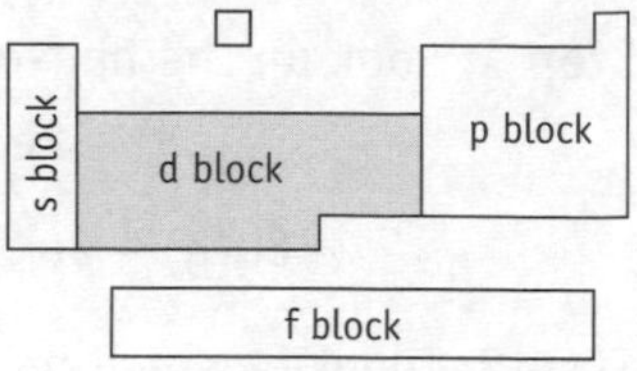

Electronic arrangements

The properties of the ten elements from scandium to zinc are remarkably similar. This is because, as atomic number increases, each additional electron enters the 3d subshell. The 4s subshell has slightly lower energy than the (inner) 3d subshell and is filled first. Note the unusual electron arrangements for Cr and Cu; this is due to the additional stability associated with a half-full and a completely full 3d subshell.

When d block elements react to form ions, the 4s electrons are the first to be lost.

For example: Fe [Ar] $3d^6 4s^2$ Fe^{2+} [Ar] $3d^6 4s^0$ Fe^{3+} [Ar] $3d^5 4s^0$

Scandium and zinc don't display the chemical properties associated with the transition metals because their ions (Sc^{3+} and Zn^{2+}) have electron arrangements $3d^0$ and $3d^{10}$ respectively.

Physical properties

In comparison with the s block metals, d block metals are denser, have higher melting and boiling points and are also good conductors of heat and electricity. They are hard and durable, with high tensile strength and good mechanical properties. This makes them ideal for a wide range of uses, both as pure metals and in alloys.

Chemical properties

The transition metals have four important chemical properties, all of which relate directly to the electronic arrangements of the elements or their ions. These are:

• Variable oxidation states

This is because the gaps between successive ionisation enthalpies in the 3d and 4s subshells are relatively small, so multiple electron loss is possible. In the lower oxidation states, the elements exist as simple ions (e.g. Cu^{2+}, Cr^{3+}, Fe^{2+}, Fe^{3+}), while in the higher oxidation states they are covalently bonded to electronegative elements such as oxygen or fluorine, forming anions (e.g. $Cr_2O_7^{2-}$, MnO_4^-, VO_3^-). Compounds containing metals in high oxidation states tend to be oxidising agents whereas compounds with metals in low oxidation states are often reducing agents.

Outline of the Periodic Table

A **transition metal** is an element that forms at least one ion with a **partially filled subshell** of d electrons.

Electronic arrangements of the d block elements in Period 4:
Sc [Ar] $3d^1 4s^2$
Ti [Ar] $3d^2 4s^2$
V   [Ar] $3d^3 4s^2$
Cr [Ar] $3d^5 4s^1$
Mn [Ar] $3d^5 4s^2$
Fe [Ar] $3d^6 4s^2$
Co [Ar] $3d^7 4s^2$
Ni [Ar] $3d^8 4s^2$
Cu [Ar] $3d^{10} 4s^1$
Zn [Ar] $3d^{10} 4s^2$
[Ar] = $1s^2 2s^2 2p^6 3s^2 3p^6$

Common oxidation states of manganese:
+2 Mn^{2+}
+4 MnO_2
+7 MnO_4^-

• Formation of coloured ions

Transition metal compounds show many different colours. Electron transitions occur within the 3d subshell when visible light is absorbed – this can only happen in ions that have a partially filled 3d subshell. A fuller explanation of why transition metal ions appear coloured is given on page 53.

▶ Some common colours:
$Cu^{2+}(aq)$ blue
$Fe^{2+}(aq)$ green
$Fe^{3+}(aq)$ yellow
$Co^{2+}(aq)$ pink

• Formation of complexes

Transition metals are able to form complexes because their 3d orbitals can accommodate the electrons donated by the ligands. For further details about complexes see pages 36–37.

• Catalytic activity

Many catalysts are made from transition metals or their compounds. The metals can act as **heterogeneous** catalysts, providing a surface onto which gaseous reactant molecules are adsorbed. Weak interactions between these and the 3d and 4s electrons of the transition metal keep the molecules in place while bonds are broken and formed.

▶ Some important heterogeneous catalysts are Fe used in the Haber process, V_2O_5 used in the contact process, Ni used in hydrogenation, and Pt/Rh used in catalytic converters.

Transition metal ions can also act as **homogeneous** catalysts. For example, Fe^{2+} ions catalyse the oxidation of iodide ions by peroxodisulphate ions.

Reactions with sodium hydroxide and aqueous ammonia

Most transition metal ions form precipitates with sodium hydroxide solution, e.g.

$$Fe^{2+}(aq) + 2OH^-(aq) \rightarrow Fe(OH)_2(s)$$
dark green solid

Similar precipitation reactions occur if aqueous ammonia is added to aqueous transition metal ions. Excess aqueous ammonia causes a soluble ammine complex to form, e.g.

$$[Cu(H_2O)_6]^{2+}(aq) + 2OH^-(aq) \rightarrow [Cu(OH)_2(H_2O)_4](s) + 2H_2O(l)$$
pale blue solid

▶ The colour of the hydroxide can be used to determine the cation present in the initial solution. Blue indicates the presence of Cu^{2+}, green Fe^{2+} and orange Fe^{3+}. Be sure you can write ion equations for each of these.

then

$$[Cu(OH)_2(H_2O)_4](s) + 6NH_3(aq) \rightarrow [Cu(NH_3)_6]^{2+}(aq) + 4H_2O(l) + 2OH^-(aq)$$
deep blue solution

▶ The OH⁻ is formed in the reaction $NH_3(aq) + H_2O(l) \rightleftharpoons NH_4^+(aq) + OH^-(aq)$

❓ *Quick check questions*

1 Write the electron arrangement of the following transition metal ions:
   (a) Ti^{2+} **(b)** Cr^{3+} **(c)** V^{3+}.

2 Give the oxidation state of the transition metal in the following compounds or ions: **(a)** CrO_4^{2-} **(b)** MnO_2 **(c)** VO_2^+.

3 Copper can form Cu^+ and Cu^{2+} ions. Is copper a transition metal?

4 A solution of sodium potassium tartrate is oxidised by hydrogen peroxide solution to carbon dioxide, water and methanoate ions. The reaction is catalysed by $Co^{2+}(aq)$ ions.

   (a) Is the Co^{2+} a heterogeneous or homogeneous catalyst?

   (b) Suggest what happens to the Co^{2+} ions during the course of the reaction.

Complex formation
Chemical Ideas 11.6

In a complex, a central metal atom or ion is surrounded by ligands. Ligands are molecules or anions with one or more lone pairs of electrons. An example of a complex ion is $[Cu(H_2O)_6]^{2+}$, shown on the right.

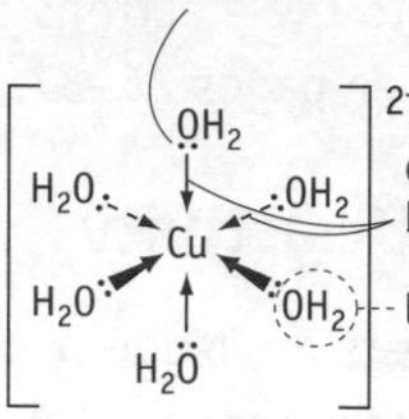

The ligands form **dative covalent bonds** with the central metal. The number of bonds between the central metal and the ligands is called the **coordination number** of the central metal. The complex ion shown on the right has a coordination number of 6.

lone pair of electrons on oxygen forms the dative bond

dative covalent bonds

ligand

Shapes of complexes

Complexes with coordination number 6 are generally **octahedral** in shape. Those with coordination number 4 are usually **tetrahedral** but can be **square planar** while those with coordination number 2 are **linear**. For example:

> If a complex has an overall charge, it is called a **complex ion**. Always draw a square bracket round a complex ion and write the charge outside the bracket.

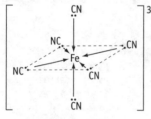

shape: octahedral
coordination number: 6

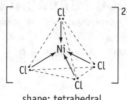

shape: tetrahedral
coordination number: 4

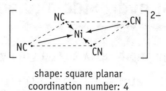

shape: square planar
coordination number: 4

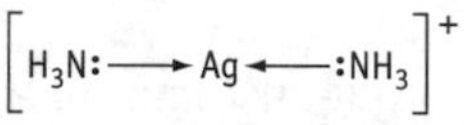

$$\left[H_3N: \longrightarrow Ag \longleftarrow :NH_3 \right]^+$$

shape: linear
coordination number: 2

Types of ligands

Ligands can form **one** bond to the central metal (**mono**dentate), **two** bonds (**bi**dentate) or **many** bonds (**poly**dentate). Here are the names and formulae of some common ligands:

> A single edta^{4-} ion can form a hexadentate complex by wrapping itself around the central transition metal ion. This is known as chelation.

Molecule/ion	Formula	Name of ligand	Type of ligand
water	H_2O	aqua	monodentate
ammonia	NH_3	ammine	monodentate
chloride ion	Cl^-	chloro	monodentate
cyanide ion	CN^-	cyano	monodentate
hydroxide ion	OH^-	hydroxo	monodentate
ethane-1,2-diamine	$H_2NCH_2CH_2NH_2$	ethane-1,2-diamine or (en)	bidentate
EDTA	You don't need to remember this formula!	edta^{4-}	hexadentate (polydentate)

Naming complexes

The name of a complex can be worked out by applying the following rules, in order:

- Write the number of each type of ligand, using the prefixes mono, di, tri, tetra, penta, hexa.
- Write the name of each ligand (in alphabetical order).
- Write the name of the central metal. Use the English name if the overall charge on the complex is positive or neutral and the Latinised name if it is negative.
- Write the oxidation number of the central metal, in brackets.

Metal	Latinised name
Cu	cuprate
V	vanadate
Ti	titanate
Ag	argentate
Zn	zincate
Pb	plumbate
Cr	chromate

For example:

$[CuCl_4]^{2-}$ tetrachlorocuprate(II) ion
$[Cu(NH_3)_4(H_2O)_2]^{2+}$ tetraamminediaquacopper(II) ion

Stability of complexes

Some ligands bond more strongly to the central metal than others. For example, if concentrated hydrochloric acid is added dropwise to a solution of copper sulphate, chloride ligands replace water ligands and the solution changes colour from blue to green. This type of reaction is called a **ligand exchange** reaction.

$[Cu(H_2O)_6]^{2+} + 4Cl^-(aq) \rightleftharpoons [CuCl_4]^{2-}(aq) + 6H_2O(aq)$
blue   green

▶ Ligand exchange reactions are also called ligand substitution reactions.

Stability constants provide a numerical way of comparing the stability of different complexes. The stability constant for the reaction above is written:

$$K_{stab} = \frac{[[CuCl_4]^{2-}(aq)]}{[[Cu(H_2O)_6]^{2+}(aq)][Cl^-(aq)]^4} = 3.98 \times 10^5 \ mol^{-4}dm^{12}$$

The larger the value of the stability constant, K_{stab}, the more stable the complex is.

Because the values of stability constants can be very high, they are often quoted as logarithms (log K_{stab}).

▶ A stability constant, written K_{stab}, is another type of equilibrium constant. The only thing that affects its value is a change in temperature. Notice that the concentration of water, $[H_2O]$, does **not** appear in the K_{stab} expression.

? Quick check questions

1. Explain what is meant by the following terms:
   (a) ligand (b) complex ion
   (c) coordination number (d) dative covalent bond.

2. Draw and name the following complexes:
   (a) $[Cr(H_2O)_6]^{3+}$ (b) $[CoCl_4]^{2-}$ (c) $[Fe(OH)_2(H_2O)_4]^+$

3. The equation for a ligand exchange reaction is given below:
   $$[Cu(H_2O)_6]^{2+}(aq) + 4NH_3(aq) \rightleftharpoons [Cu(NH_3)_4(H_2O)_2]^{2+}(aq) + 4H_2O(l)$$
   (a) Write an expression for the stability constant, K_{stab}, for the reaction.
   (b) log K_{stab} for the above reaction is 12. Compare this value with that for the reaction shown in the text above and use the data to explain which complex is the more stable.

Aspects of Agriculture (AA)

Successfully producing crops depends on many factors. The chemistry of some of these, such as soil composition, nutrients and pesticides, are studied in this unit. CI refers to sections in your Chemical Ideas textbook.

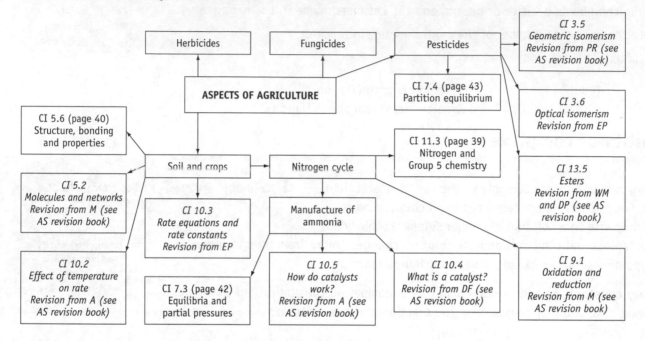

Several parts of this unit do not fit into the main Chemical Ideas sections but are part of the materials you may be examined on. These are:

Improving food production (Storyline AA1)

Plants need the right conditions to grow well:

- The main nutrients – N as nitrate(V) ions (NO_3^-), P as phosphate ions (PO_4^{3-}) and K as potassium ions (K^+) – have to be available in solution. Adding fertiliser is one way of increasing plant growth, but the amounts and timing of the application are critical.

- Plants have an optimum pH range. Farmers increase the pH of soil by adding lime.

- Controlling pests by the use of pesticides can increase yields significantly.

Manufacture of ammonia (Haber Process) (Storyline AA3)

The raw materials for this process are air and natural gas. From these a feedstock containing the reactants, nitrogen and hydrogen, in the ratio 1:3, is produced.

▶ $N_2(g) + 3H_2(g) \rightleftharpoons 2NH_3(g)$

The usual conditions for this process are: iron catalyst, temperature 450°C and 200 atm pressure. These conditions are a compromise to produce reasonable yields with an acceptable rate of attainment of equilibrium, e.g. increasing temperature decreases yields but increases rates; increasing pressure increases both yields and rates but is expensive, because of the high-pressure plant and equipment needed.

Nitrogen and Group 5
Chemical Ideas 11.3

This group is in the middle of the p block, with all the elements having electron configurations ending with p^3 (e.g. nitrogen $1s^2 2s^2 2p^3$).

All these elements can have oxidation states of +3 or +5 in compounds. The +5 oxidation state is more common for non-metal compounds.

Nitrogen gas consists of diatomic molecules ($:N{\equiv}N:$), which are very unreactive because of the extremely high activation energy needed to start breaking bonds. Phosphorus, on the other hand, exists as P_4 molecules, with single bonds between all atoms, and is very reactive.

Nitrogen	N	
Phosphorus	P	
Arsenic	As	increasing metallic nature
Antimony	Sb	
Bismuth	Bi	

Nitrogen cycle

This cycle involves natural redox reactions for interconversions between nitrogen species in the soil and/or the atmosphere.

Nitrogen species	Formula	Oxidation state	Action producing the species
Nitrogen gas	$N_2(g)$	0	Denitrifying bacteria in soil
Nitrate(V) ion	$NO_3^-(aq)$	+5	Nitrifying bacteria in soil
Nitrate(III) ion	$NO_2^-(aq)$	+3	Nitrifying bacteria in soil
Ammonium ion	$NH_4^+(aq)$	−3	Root nodules in legumes Bacteria and microorganisms in soil
Dinitrogen oxide	$N_2O(g)$	+1	Denitrifying bacteria in soil
Nitrogen monoxide [nitrogen(II) oxide]	$NO(g)$	+2	Car engines, thunderstorms, denitrifying bacteria in soil
Nitrogen dioxide [nitrogen(IV) oxide]	$NO_2(g)$	+4	Oxidation of NO in atmosphere

Shapes of common nitrogen compounds

The bonding in some common nitrogen compounds is shown below:

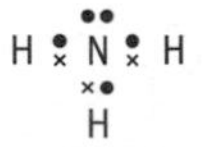

ammonia (pyramidal) 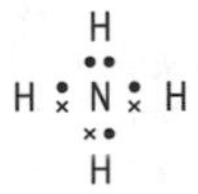   ammonium ion (tetrahedral) 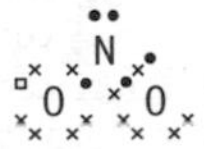   nitrate(III) 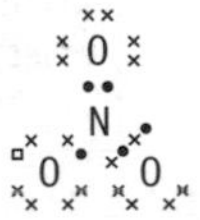   nitrate(V) (triangular planar)

? Quick check questions

1 In addition to those in the above table, what two processes remove nutrients, such as nitrogen compounds, from the soil?

2 Which type of soil bacteria carries out oxidation? Explain.

Bonding, structure and properties
Chemical Ideas 5.6

The properties of materials are determined by:

- The types of particles in that material, e.g. atoms, ions or molecules.
- The bonding in that material, e.g. covalent, ionic, metallic, intermolecular forces.
- The structure present in that material, e.g. giant lattice, molecular or macromolecular.

	GIANT LATTICE			COVALENT MOLECULAR	
	Ionic	Covalent network	Metallic	Simple molecular	Macromolecular
What substances have this type of structure?	compounds of metals with non-metals	some elements in Group 4 and some of their compounds	metals	some non-metal elements and some non-metal/ non-metal compounds	polymers
Examples	sodium chloride, NaCl; calcium oxide, CaO	diamond C; graphite, C; silica, SiO_2	sodium, Na; copper, Cu; iron, Fe	carbon dioxide CO_2, chlorine Cl_2, water H_2O	poly(ethene), nylon, proteins, DNA
What type of particles does it contain?	ions	atoms	positive ions surrounded by delocalised electrons	small molecules	long-chain molecules
How are the particles bonded together?	strong ionic bonds; attraction between oppositely charged ions	strong covalent bonds	strong metallic bonds; attraction of atoms' nuclei for delocalised electrons	weak intermolecular bonds between molecules; strong covalent bonds between the atoms within each molecule	weak intermolecular bonds between molecules; strong covalent bonds between the atoms within each molecule
What are the typical properties?					
Melting point and boiling point	high	very high	generally high	low	moderate (often decompose on heating)
Hardness	hard but brittle	very hard (if three-dimensional)	hard but malleable	soft	variable; many are soft but often flexible
Electrical conductivity	conduct when molten or dissolved in water; electrolytes	do not normally conduct	conduct when solid or liquid	do not conduct	do not normally conduct
Solubility in water	often soluble	insoluble	insoluble (but some react)	usually insoluble, unless molecules contain groups which can hydrogen bond with water	usually insoluble
Solubility in non-polar solvents (e.g. hexane)	generally insoluble	insoluble	insoluble	usually soluble	sometimes soluble

The periodic table allows trends in bonding, structure and properties to be recognised. For example, for the first 20 elements of the periodic table:

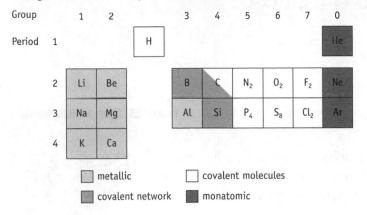

All of the elements in Period 3 react with oxygen to form oxides and with chlorine to form chlorides:

Group	1	2	3	4	5	6	7	0
Element	Na	Mg	Al	Si	P	S	Cl	Ar
Structure of element	←——— metallic lattice ———→			covalent network	←—— covalent molecules ——→			monatomic
Reaction of element with water	vigorous to form hydroxide and hydrogen	with steam gives hydroxide and hydrogen	X	X	X	X	dissolves and reacts to give acidic solution (HCl + HClO)	X
Structure of oxide	←—— ionic lattice ——→		ionic/ covalent layer lattice	covalent network	←——— covalent molecules ———→			
Formula of oxide	Na_2O	MgO	Al_2O_3	SiO_2	P_4O_{10} P_4O_6	SO_3 SO_2	Cl_2O_7 Cl_2O	—
Acid-base character of oxide	←—— basic ——→		amphoteric	←——————— acidic ———————→				
Structure of chloride	←—— ionic lattice ——→		←———————— covalent molecules ————————→					
Formula of chloride	NaCl	$MgCl_2$	$AlCl_3$	$SiCl_4$	PCl_5 PCl_3	S_2Cl_2	Cl_2	—
What happens when the chloride is put into water?	←—— dissolves ——→ to form neutral solution	to form acidic solution	←———— reacts with water ————→ producing fumes of hydrogen chloride and an acidic solution				some of the chlorine reacts with water to form an acidic solution	

? Quick check questions

1 Use the tables in this section to give a structure for each of the following:

   (a) vanadium   (b) xenon   (c) cotton   (d) potassium iodide

   (e) propan-1-ol   (f) steel   (g) glass   (h) polyester

   (i) lead nitrate   (j) silicon carbide

2 For each substance listed in question 1, give the following properties:

   (a) state at room temperature (solid, liquid or gas)

   (b) solubility in water (soluble or insoluble)

   (c) electrical conductivity (high or low).

Equilibria and partial pressures

Chemical Ideas 7.3

For reactions in the gas phase, partial pressures (the pressure exerted by a gas in a mixture of gases) can be used, instead of concentrations, in an equilibrium expression. For example, in the reaction between hydrogen gas and iodine vapour:

$$H_2(g) + I_2(g) \rightleftharpoons 2HI(g) \qquad K_p = \frac{p_{HI}^2}{p_{H_2} \times p_{I_2}}$$

> Like all equilibrium constants, K_p is a constant and only changes with temperature. Changes in pressure or using a catalyst do not alter the value of K_p.

Also, the sum of the partial pressures is equal to the total pressure:

$$\text{Total pressure } P = p_{H_2} + p_{I_2} + p_{HI}$$

Worked example

$$N_2(g) + 3H_2(g) \rightleftharpoons 2NH_3(g)$$

In the Haber process (equation above), nitrogen and hydrogen gases are mixed in a ratio of 1:3. This mixture is passed over the iron catalyst at 723 K and 200 atm. Assuming equilibrium is reached with 25.3% ammonia in the mixture, calculate K_p.

At equilibrium the % ammonia = 25.3%

At equilibrium the % nitrogen = 1/4 × 74.7 = 18.7%   ⎫
                                                     ⎬ These % total 100%
At equilibrium the % hydrogen = 3/4 × 74.7 = 56.0%   ⎭

At equilibrium the total pressure = 200 atm = $p_{H_2} + p_{N_2} + p_{NH_3}$

Partial pressure of ammonia $(p_{NH_3}) = \dfrac{25.3}{100} \times 200 = 50.6$ atm   ⎫

Partial pressure of nitrogen $(p_{N_2}) = \dfrac{18.7}{100} \times 200 = 37.4$ atm   ⎬ These pressures total 200 atm

Partial pressure of hydrogen $(p_{H_2}) = \dfrac{56.0}{100} \times 200 = 112$ atm   ⎭

$$K_p = \frac{p_{NH_3}^2}{p_{N_2} \times p_{H_2}^3} = \frac{(50.6 \text{ atm})^2}{37.4 \text{ atm} \times (112 \text{ atm})^3}$$

$$K_p = 4.9 \times 10^{-5} \text{ atm}^{-2}$$

> The units for K_p need to be calculated for every example. Sometimes K_p can have units, as in this case:
> $$\text{units} = \frac{\text{atm}^2}{\text{atm} \times \text{atm}^3}$$
> $$= \text{atm}^{-2}$$

? Quick check questions

1 Give an expression for K_p and its units for the reaction between methane and steam:

$$CH_4(g) + H_2O(g) \rightleftharpoons CO(g) + 3H_2(g)$$

2 Calculate a value for K_p if the equilibrium partial pressures in the above reaction are methane 0.20 atm, hydrogen 0.49 atm, carbon monoxide 0.20 atm, and the conditions are 120°C and 1.05 atm.

> Remember significant figures.

Partition equilibrium
Chemical Ideas 7.4 and Storyline AA4

Partition equilibria relate to a substance dissolving in, or distributing itself between, different phases and coming to equilibrium.

The most common type is a solid dissolving in two immiscible liquids.

For example, iodine added to a mixture of water and tetrachloromethane, and then left to reach equilibrium:

$$I_2(aq) \rightleftharpoons I_2(organic)$$

At equilibrium, the partition coefficient $K_{ow} = \dfrac{[I_2(organic)]}{[I_2(aq)]}$

> ▶ Phases – different physical states, e.g. solid, liquid or gas. Immiscible – does not mix and so forms two layers.

Partition coefficients are particularly important for insecticides, where K_{ow} needs to be large so as to dissolve in the fatty layers of the insect in sufficient quantities to kill it, even though the aqueous spray is very dilute. Unfortunately, if the insecticide is not biodegradable (e.g. DDT), it will build up in the food chain and cause problems. All modern insecticides are biodegradable (e.g. pyrethroids) for this reason.

> ▶ As the units of concentration in K_{ow} cancel out, it does not matter what units are used, as long as the units for each layer are the same.

Chemists will use a promising compound that is discovered to be active against pests. They will try to improve its effectiveness by making systematic changes to the structure, continuing to test and work out the best substitutions in various parts of the molecule. As well as considering potency with respect to the target group of insects, the chemists will need to consider environmental factors such as how easily the product leaches into water supplies, or how far it will accumulate in food chains.

❓ Quick check questions

1 Calculate the partition coefficient of iodine between tetrachloromethane and water (K_{ow}), if at equilibrium the concentration of iodine in water is 0.202 mol dm^{-3} and in tetrachloromethane is 17.3 mol dm^{-3}.

2 0.77 g of chlorine are added to a mixture of 100 cm^3 of each of tetrachloromethane and water. After shaking repeatedly, the mass of chlorine in the aqueous layer was found to be 0.07 g. Calculate the partition coefficient of chlorine between tetrachloromethane and water (K_{ow}).

3 Given that K_{ow} for DDT is 9.5×10^5 and the concentration in the aqueous layer of a mixture of octan-1-ol and water is 0.002 g dm^{-3}, find the concentration of DDT in the octan-1-ol.

Colour by Design (CD)

In this unit you are able to discover what causes colour. The unit explores the restoration of an oil painting and the associated chemistry and analytical techniques needed to carry this out. The unit also looks at the development of synthetic dyes for cloth. CI refers to sections in your Chemical Ideas textbook.

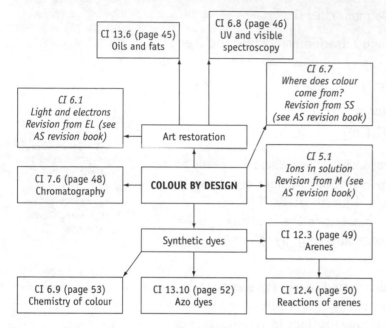

Several parts of this unit do not fit into the main Chemical Ideas sections but are part of the materials you may be examined on. These are:

- Colour changes may be associated with a range of types of chemical reactions. These include acid–base (indicators), ligand exchange, redox, precipitation and polymorphism (different crystalline forms of a compound). **(Activity CD1)**

- Pigments need to have high colour intensities (only a small amount is needed to give a vibrant colour). They need to be fast to light (i.e. not fade) and they need to be of the correct shade for the application. **(Storyline CD2)**

- Oil paints harden by a process known as 'oxidative crosslinking'. The unsaturated side chains in the oil react with oxygen. This is followed by a polymerisation reaction in which crosslinking takes place between the chains of the triesters. The result is a vast interlocking network of triesters joined by C—C, C—O—C and C—O—O—C bridges. These reactions are much more rapid in the presence of oxygen and light, suggesting a free radical mechanism. **(Activity CD4.2)**

- When dyes attach themselves to fibres, they do so using a variety of bonds and intermolecular forces. In the case of dyes that attach successfully to cotton these interactions will be hydrogen bonds. In the case of dyes that attach to protein fibres, e.g. wool or silk, these interactions are most likely to be ionic interactions. Other dyes are attached to a fabric using a mordant. This mordant (usually a metal hydroxide) precipitates onto the surface of the fabric and allows a complex to form between the fabric and the dye. Some synthetic dyes have been produced which form strong covalent bonds with the fabric. These are more difficult to remove by washing and do not fade as quickly. **(Storyline CD7)**

Oils and fats

Chemical Ideas 13.6

Oils and fats are naturally occurring **triesters** of propane-1,2,3-triol (glycerol) and long chain carboxylic acids (fatty acids). Each of the three ester groups can be from the same fatty acid or they can be different (mixed triester).

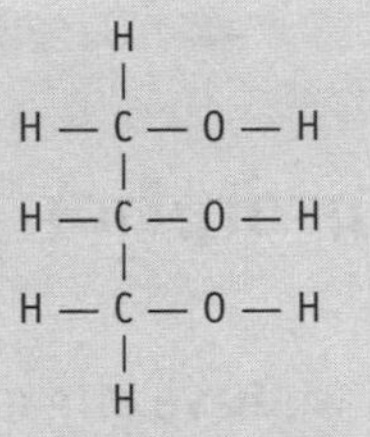

Fatty acids

Fatty acids are the carboxylic acids in fats and oils. They have an even number of carbons (typically 16 or 18) in their unbranched carbon chain. These chains can be either saturated (containing single C—C bonds) or unsaturated (contain some double C=C bonds). A typical saturated and unsaturated fatty acid are shown below:

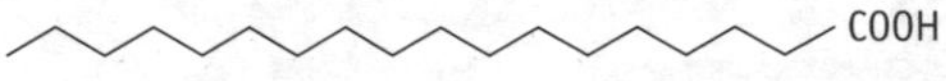

COOH   stearic acid
• saturated
• no double bonds

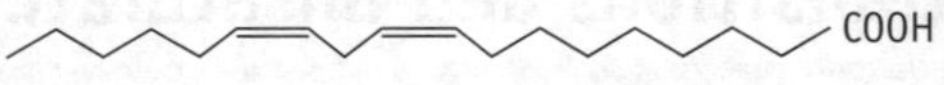

COOH   linoleic acid
• unsaturated
• two double bonds

Any natural oil or fat contains a mixture of triesters. Triesters from largely saturated fatty acids are solids or fats, as there is better packing of the molecules resulting in stronger intermolecular forces. Triesters from largely unsaturated fatty acids are liquids or oils.

Hydrolysis (saponification) of esters

Any natural oil or fat can be broken down into the sodium salt of the fatty acid (soap) and glycerol by heating with dilute sodium hydroxide:

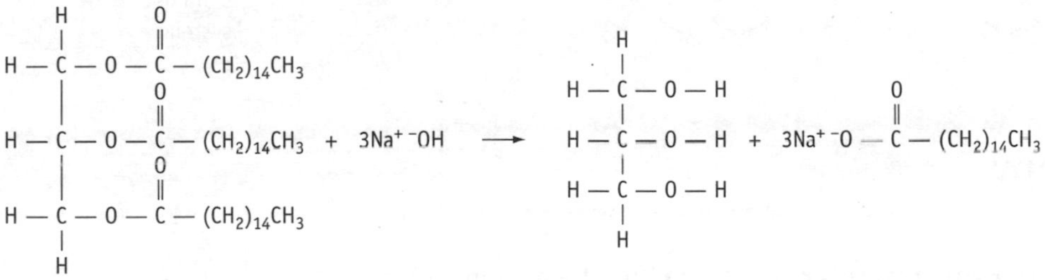

$$H-\overset{\overset{\displaystyle H}{|}}{C}-O-\overset{\overset{\displaystyle O}{||}}{C}-(CH_2)_{14}CH_3$$

$$H-\overset{\overset{\displaystyle |}{C}}{C}-O-\overset{\overset{\displaystyle O}{||}}{C}-(CH_2)_{14}CH_3 \; + \; 3Na^+\,{}^-OH \longrightarrow$$

$$H-\overset{\overset{\displaystyle |}{C}}{C}-O-\overset{\overset{\displaystyle O}{||}}{C}-(CH_2)_{14}CH_3$$

$$\overset{\displaystyle |}{H}$$

triester formed from glycerol and palmitic acid

$$H-\overset{\overset{\displaystyle H}{|}}{C}-O-H$$
$$H-\overset{\displaystyle |}{C}-O-H \; + \; 3Na^+\,{}^-O-\overset{\overset{\displaystyle O}{||}}{C}-(CH_2)_{14}CH_3$$
$$H-\overset{\displaystyle |}{C}-O-H$$
$$\overset{\displaystyle |}{H}$$

glycerol   sodium palmitate

Hydrogenation (reduction or addition)

Addition of hydrogen to unsaturated oils, using a **nickel** catalyst and the correct conditions, produces a more saturated solid fat, which can be used in the manufacture of margarine. Not all the C=C double bonds are hydrogenated so as to give a spreadable fat, which is still polyunsaturated.

Sidebar

$$H-\overset{\overset{\displaystyle H}{|}}{C}-O-H$$
$$H-\overset{\displaystyle |}{C}-O-H$$
$$H-\overset{\displaystyle |}{C}-O-H$$
$$\overset{\displaystyle |}{H}$$

propane-1,2,3-triol (glycerol)

Triesters are compounds containing three ester groups.

If the free acid is required it can be released by treating the sodium salt with dilute hydrochloric acid.

? Quick check questions

1 Give the skeletal formula for the fat made from stearic acid, $CH_3(CH_2)_{16}COOH$.

2 Why are saturated fats solids?

3 How would you decide if an oil was linseed oil or not?

Ultraviolet and visible spectroscopy
Chemical Ideas 6.8

Absorption of ultraviolet and visible light

As we have already seen on page 30, if visible light falls on a coloured solution, some wavelengths are absorbed and some are transmitted. What we see are the transmitted wavelengths (colours). Many molecules also absorb ultraviolet (UV) radiation. Because our eyes don't detect UV light, a molecule that absorbs only UV light (and transmits all visible wavelengths) appears colourless.

> UV radiation has wavelengths between 280 and 400 nm. It has higher energy than visible radiation.

Electronic transitions and unsaturated molecules

Absorption of UV or visible light causes **electronic transitions** in molecules – electrons move to higher energy levels and the molecule becomes **'excited'**. Unsaturated molecules (i.e. molecules with C=C double bonds or benzene rings) and those with **conjugated** systems often absorb UV and visible light. The **delocalised** electrons in these systems require slightly less energy to become excited compared with electrons in single bonds.

> A conjugated system has alternate C=C and C—C bonds. Benzene rings and N=N double bonds can also be part of conjugated systems.

Carotene is an example of a large conjugated molecule:

UV-visible spectroscopy

In UV and visible spectrometry, the **spectrometer** scans a range of wavelengths of both UV and visible light. The instrument produces a **spectrum** such as the one shown below for carotene:

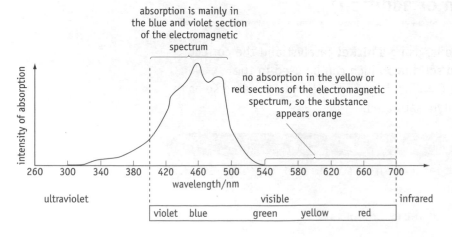

Notice the following features of the spectrum:

- The x-axis shows wavelength, measured in nanometres (nm).
- The y-axis shows intensity of absorption; usually there are no units.
- Unlike an infrared spectrum (see page 6), the trace from a UV-visible spectrum is much broader. It is the overall shape of the spectrum that is important rather than individual peaks.

> ◗ 1 nm = 1×10^{-9} m

Interpreting the spectrum

To identify the colour of the solution being tested, find the wavelength at which absorption is greatest – this is known as λ_{max}.

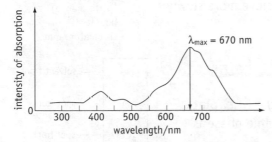

For example, in the spectrum above, λ_{max} is at 670 nm. This is the wavelength of red light. The colour of the solution will be the complementary colour of red, which is blue/green. Using a colour wheel will help you find the complementary colour (see page 30).

> ◗ Absorption is more intense and the wavelength of λ_{max} is longer for organic molecules with large delocalised systems.

Reflectance spectra

UV and visible spectrometry is also used when analysing substances that cannot be easily made into a solution. UV and visible light is shone onto the surface of the sample (pigment M) and any reflected light is collected and analysed – the result is a **reflectance spectrum**. One area where this technique is used is in the analysis of pigments from old paintings.

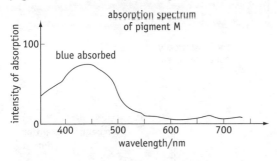
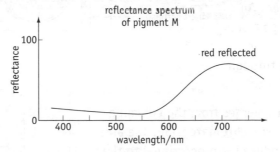

Compare the absorption and reflectance spectra of the pigment above. Where there is a peak in the absorption spectrum there is a trough in the reflectance spectrum, and vice versa.

? *Quick check question*

1 Study the absorption and reflectance spectra of pigment M above. Suggest, with reasons, the likely colour of the pigment.

Chromatography
Chemical Ideas 7.6

Chromatography is a method of separating and identifying the components of a mixture, e.g. the components that make up oils used in painting.

All types of chromatography depend on the equilibrium set up when the components of the mixture distribute themselves between the **stationary phase** and the **mobile phase**. Components with a higher affinity for the stationary phase move more slowly than those with lower affinity.

Thin layer chromatography

In thin layer chromatography (t.l.c.), the stationary phase (called the **t.l.c. plate**) is silica, spread in a thin layer onto a plastic backing sheet. The mobile phase is the solvent.

The distance that a component travels relative to the solvent front is called its **R_f value**.

Gas-liquid chromatography

In gas-liquid chromatography (g.l.c.), the stationary phase is a non-volatile liquid coated onto the surface of finely divided solid particles. This material is packed inside a long thin **column**, which is coiled inside an oven. An unreactive **'carrier gas'** acts as the mobile phase and carries the mixture through the column.

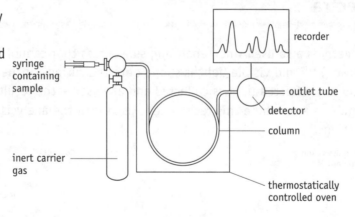

As each component emerges from the column a peak is recorded on the **chromatogram**. The area under each peak produced is proportional to the amount of that component in the mixture. The time that a component takes to emerge is called the **retention time**.

> There are many different types of chromatography – paper, thin layer, column, gas-liquid and high-pressure (performance) liquid chromatography.

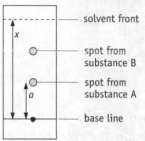

R_f value for substance A $= \dfrac{a}{x}$

> Full details of the step-by-step process for carrying out t.l.c. are given on page 76 of this book.

> Nitrogen and noble gases are used as carrier gases.

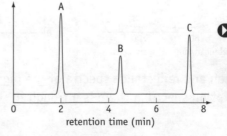

Compound	Retention time
A	2 min
B	4.5 min
C	7.4 min

? Quick check questions

1 What is the R_f value for spot a on the t.l.c. plate shown above?

2 Why are the methyl esters of fatty acids used to run g.l.c. traces rather than the acids themselves?

Arenes

Chemical Ideas 12.3

The simplest arene is benzene, C_6H_6, which has a flat hexagonal structure with a bond angle of 120°:

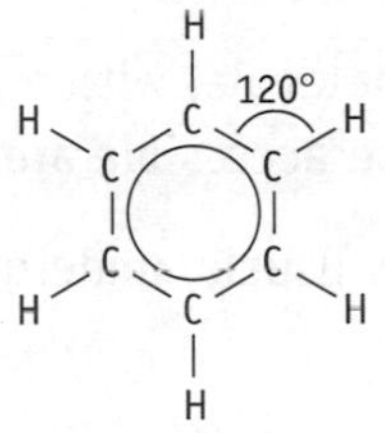

Arenes are hydrocarbons that contain benzene rings. Their names always end in –ene, meaning that they are unsaturated (as in alk<u>ene</u>).

All carbon–carbon bond lengths are the same (0.139 nm), with a value between that of a single and double bond. This means that one electron from each carbon is delocalised over the whole ring:

This arrangement is much more stable than an alternate single/double bond arrangement, as demonstrated by the enthalpy change diagram below:

regions of higher electron density above and below the benzene ring

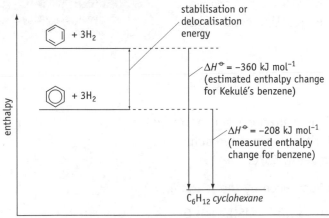

C_6H_5 is called phenyl

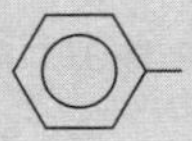

Other examples of arenes (aromatic hydrocarbons) include:

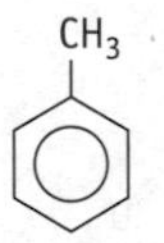

methylbenzene

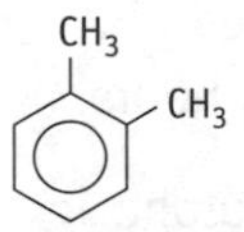

1,2-dimethylbenzene

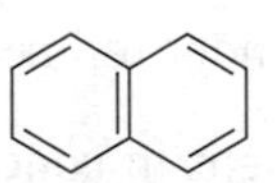

naphthalene ($C_{10}H_8$)

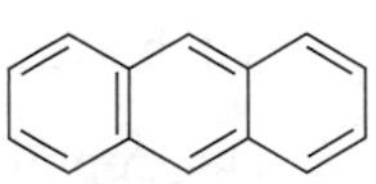

anthracene ($C_{14}H_{10}$)

When naming substituted benzenes, use the lowest numbers possible and list the groups in alphabetical order.

None of these structures contain a C=C bond and therefore they do not react with bromine water.

 Quick check questions

1 Draw the structures of **(a)** 1,3,5-trimethylbenzene **(b)** phenylethene.

2 How many electrons are delocalised in **(a)** benzene **(b)** naphthalene?

3 According to the diagram above, what is the value of the delocalisation energy for benzene?

Reactions of arenes
Chemical Ideas 12.4

Benzene is a planar molecule, with areas of delocalised electron density above and below the plane. These accessible areas of electron density will attract electrophiles.

Unlike alkenes, arenes usually undergo substitution reactions and very rarely addition reactions.

Substitution reactions keep the delocalised electron structure, which makes arenes more stable than alkenes, whereas addition would destroy this.

Consequently the vast majority of reactions are **electrophilic substitutions**.

A general equation for these reactions is:

$$R\text{-}X + E^+ \rightarrow R\text{-}E + X^+$$

R is an aryl group

X is the leaving group

E^+ is the electrophile

You need to know about the following examples. Learn the reaction conditions and the reacting electrophile in each case.

> ◆ Benzene is usually represented by
>
>
>
> in equations.

> ◆ Electrophiles are positive ions or molecules with areas of partial positive charge on one of the atoms that will be attracted to a negatively charged region.

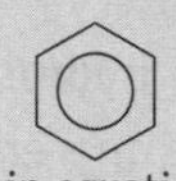

Nitration

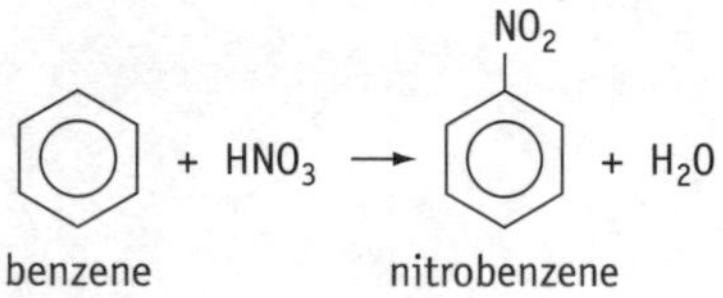

benzene + HNO$_3$ → nitrobenzene (NO$_2$) + H$_2$O

Reagents/conditions: benzene + concentrated nitric acid mixed with concentrated sulphuric acid. Temperature below 55°C.

Electrophile : NO_2^+ produced by the reaction:

$$HNO_3 + 2H_2SO_4 \rightarrow NO_2^+ + HSO_4^- + H_3O^+$$

Sulphonation

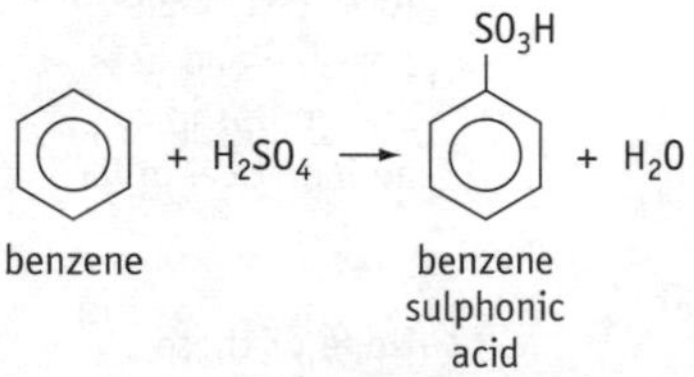

benzene + H$_2$SO$_4$ → benzene sulphonic acid (SO$_3$H) + H$_2$O

Reagents/conditions: benzene + concentrated sulphuric acid. Heat under reflux for several hours.

Electrophile: SO$_3$ (present in concentrated sulphuric acid).

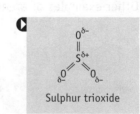

Sulphur trioxide

Chlorination

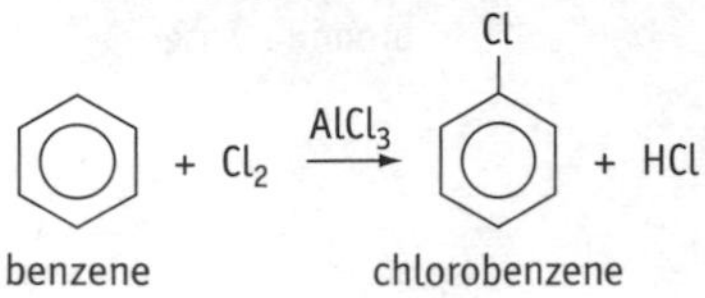

benzene + Cl$_2$ $\xrightarrow{\text{AlCl}_3}$ chlorobenzene (Cl) + HCl

Reagents/conditions: benzene + chlorine + anhydrous aluminium chloride. Room temperature.

Electrophile: $Cl^+AlCl_4^-$

Bromination

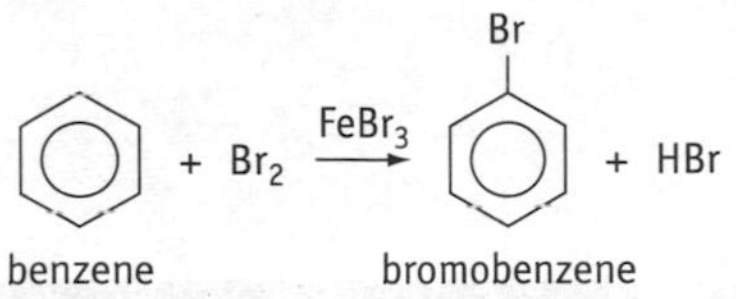

Reagents/conditions: benzene + bromine + anhydrous iron(III) bromide or iron filings. Room temperature.

Electrophile: $Br^+FeBr_4^-$

Alkylation

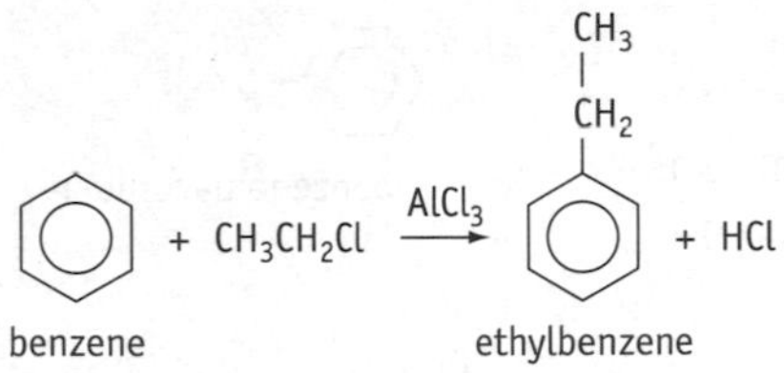

Alkylation reactions are useful in organic synthesis because they introduce an alkyl side chain onto an arene ring, allowing potential to modify the structure further.

Reagents/conditions: benzene + chloroalkane + anhydrous aluminium chloride. Heat under reflux.

Electrophile: $CH_3CH_2^{\delta+}$—Cl—$AlCl_3^{\delta-}$ or simply $CH_3CH_2^+AlCl_4^-$

> These alkylation and acylation reactions are known as Friedel Crafts reactions.

Acylation

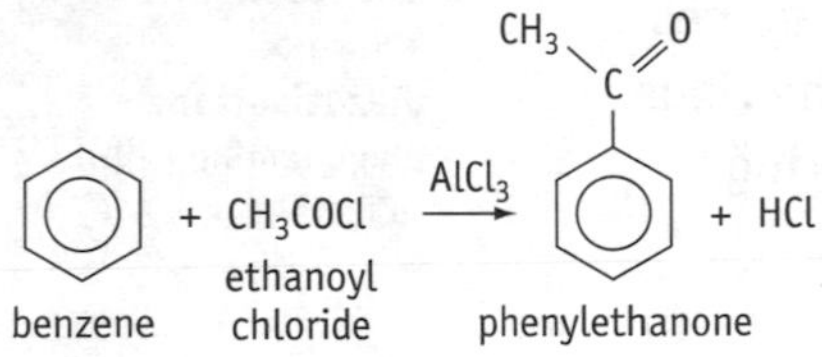

Acylation is useful in organic synthesis because it introduces an alkyl side chain with a functional group that can then be further modified.

Reagents/conditions: benzene + acyl chloride + anhydrous aluminium chloride. Heat under reflux.

Electrophile: $CH_3CO^+AlCl_4^-$

Electrophic substitution mechanism

All of the above reactions involve the electrophile attacking the electron-rich arene ring and forming a new bond with the ring. In order to regain the delocalised system one of the original ring hydrogen atoms is lost as a proton.

? Quick check questions

1 Nitration of benzene gives 1,3-dinitrobenzene – write an equation, including all reagents and reaction conditions.

2 Chlorine and bromine are weak electrophiles. How is their strength as an electrophile increased?

3 What reagents and conditions are needed to make 2-bromo-1,3-dimethylbenzene in two stages from benzene? (Hint: add the methyl groups first and then the bromine.) Write equations for the two stages.

Azo dyes

Chemical Ideas 13.10

Azo compounds make excellent dyes. They are made by reacting a 'coupling agent' (see below) with a diazonium ion.

The structure and formation of diazonium ions

Diazonium ions have the general formula $R-N^+\equiv N$. Only aromatic diazonium ions are stable, and even these have to be made in solution at temperatures below 5°C. The structure of the benzenediazonium ion is given on the right.

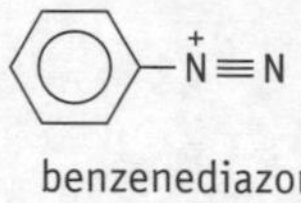

The benzenediazonium ion is made by the following reaction:

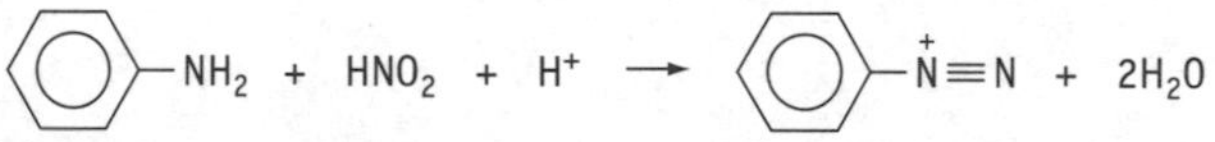

phenylamine   nitrous acid   benzenediazonium ion

The formation of azo compounds

Diazonium ions are weak electrophiles. They will attack phenols or aromatic amines (called **coupling agents**), both of which have especially electron-rich benzene rings. During the reaction, a $-N=N-$ bond is formed. The reaction is called a **coupling reaction** and the compound formed is an **azo compound**:

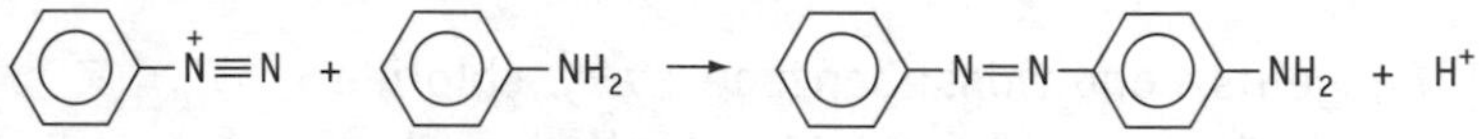

phenylamine   yellow azo compound

The uses of azo compounds

Aromatic azo compounds make excellent fade-resistant dyes. By attaching different functional groups to the chromophore, the properties of the molecule are modified, e.g.

Additional functional group(s) attached	Property modified	Example
$-SO_3H$	The solubility of the dye in water is improved	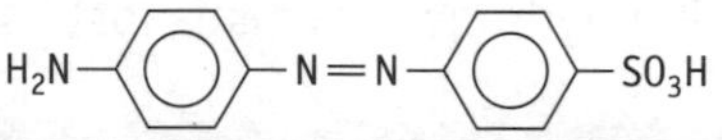
$-NH_2$ or NR_2	The colour of the dye is modified or enhanced	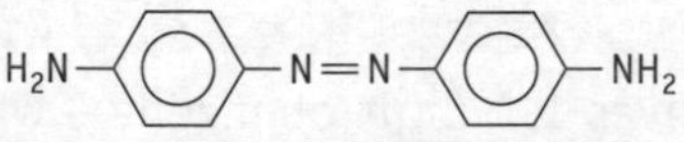

benzenediazonium ion

▶ Nitrous acid is unstable and made as needed by reacting $NaNO_2$ with HCl(aq).

▶ Learn the reactants and conditions for this reaction, known as **diazotisation**: phenylamine, dilute HCl, $NaNO_2$, < 5°C.

▶ Azo compounds have the general formula $R-N=N-R$. Compounds where the R groups are arenes are the most stable.

▶ The chromophore makes the azo dye coloured (see page 46).

? *Quick check questions*

1. Write equations for the reaction of:

   (a) dilute hydrochloric acid with sodium nitrate(III) ($NaNO_2$)

   (b) cold nitrous acid with phenylamine

   (c) benzenediazonium ion with an alkaline solution of phenol.

2. The acid–base indicator methyl orange is an azo dye. It has the structure shown below. Suggest how methyl orange could be synthesised.

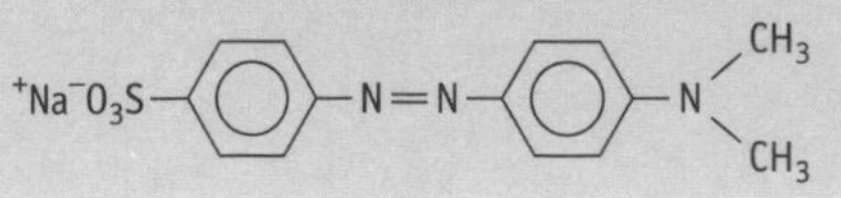

methyl orange

Chemistry of colour

Chemical Ideas 6.9

Coloured inorganic compounds

Transition metal ions are often coloured because electrons in their **d-orbitals** can be excited. When the transition metal ion is surrounded by ligands, the d-orbitals are **split** into two new energy levels.

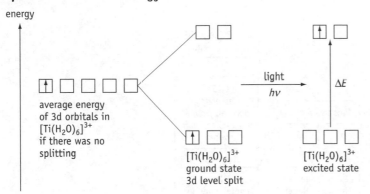

Electrons in the lower of these two new energy levels can be excited to the higher level. The excitation energy required corresponds to the absorption of visible light. Factors that affect the excitation energy (ΔE) and therefore the colour of the complex include the type of ligand; the shape of the complex – octahedral or tetrahedral; the coordination number of the complex; and the charge on the central transition metal ion.

Coloured organic compounds

The part of an organic molecule responsible for the colour is called the **chromophore**. This is an extended delocalised system of electrons containing unsaturated groups such as $C=C$, $C=O$, $N=N$ and benzene rings. Electrons in the delocalised system generally need less energy to become excited than those in single covalent bonds – this energy is available when the molecule absorbs visible light.

> ◗ In an **isolated** transition metal ion, all the d-orbitals (and any electrons in them) have the **same** energy.

> ◗ If the d subshell is empty, or completely full, then the transition metal ion is colourless.

> ◗ Examples include azo dyes (see page 52).

? *Quick check questions*

1 The complex ion $[Ni(H_2O)_6]^{2+}(aq)$ is green in colour.

   (a) Write down the electronic arrangement of the Ni^{2+} ion.

   (b) Explain, with the aid of a diagram, why this complex ion is coloured.

   (c) When the ligand edta^{4-} is added to $[Ni(H_2O)_6]^{2+}$, a blue complex ion forms. Explain why the new complex has a different colour.

2 The structure on the right is the dye 'Acid Orange 7'.

   (a) Draw a ring around the chromophore in the molecule.

   (b) Explain why Acid Orange 7 is coloured.

   (c) Suggest how the colour of Acid Orange 7 might be modified.

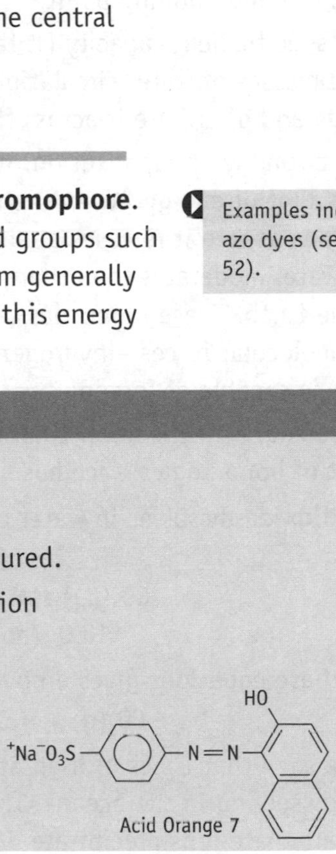

$^+Na^-O_3S$—⬡—$N=N$—

Acid Orange 7

The Oceans (O)

The vast expanses of seawater, called oceans, play an essential part in the cycling of many chemicals and also in climate control. This unit is about the fundamental chemistry that lies behind these ocean processes. CI refers to sections in your Chemical Ideas textbook.

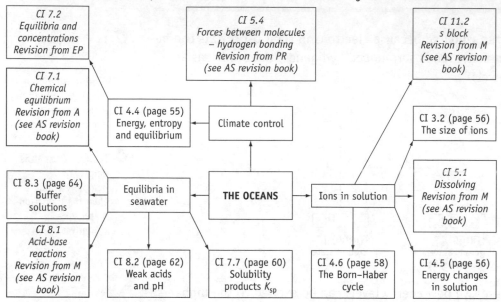

Several parts of this unit do not fit into the main Chemical Ideas sections.

Water has several unique properties:

- Relatively high boiling point.
- High specific heat capacity (it takes a lot of energy to raise its temperature). Large masses of water circulating in the oceans can carry a large amount of energy and distribute it across the world's oceans.
- High enthalpy of vaporisation. When water evaporates in equatorial regions it takes a lot of energy. When the water vapour cools towards the poles of the Earth and falls as rain it releases a large amount of energy to the surroundings. In this way water moderates the otherwise extremes of temperatures which would exist on the Earth. These properties can be explained by the strength and amount of intermolecular forces – hydrogen bonding is a strong intermolecular force and water is capable of forming two hydrogen bonds per molecule.
- The density of ice is lower than that of liquid water. The low density of ice is a result of bond angles – ice has a very open structure, with large spaces.

Carbon dioxide dissolves in water by the following route:

$$CO_2(g) \rightleftharpoons CO_2(aq)$$
$$CO_2(aq) + H_2O(l) \rightleftharpoons H^+(aq) + HCO_3^-(aq)$$
$$HCO_3^-(aq) \rightleftharpoons H^+(aq) + CO_3^{2-}(aq)$$

Adding these equations gives an overall equation:

$$CO_2(g) + H_2O(l) \rightleftharpoons 2H^+(aq) + CO_3^{2-}(aq)$$

It can be seen that times of high atmospheric carbon dioxide levels will lead to more carbon dioxide dissolving in the oceans. The oceans do not become increasingly acidic because the excess carbonate ions precipitate out as limestone (mainly calcium carbonate). Conversely, the reverse can occur if levels of atmospheric carbon dioxide fall.

Energy, entropy and equilibrium
Chemical Ideas 4.4

Entropy

- Entropy is a measure of the number of ways of arranging molecules and distributing their quanta of energy. There are more ways if the energy levels are closer together.
- Gases have higher entropies than liquids, and liquids have higher entropies than solids.
- Substances have higher entropies if their molecules contain heavier atoms or larger numbers of atoms, or if the temperature increases.

Entropy changes

The entropy change for a chemical system (ΔS_{sys}) is the difference between the entropies of the reactants and products in the equation for the reaction, e.g. ice melting.

$$H_2O(s) \rightarrow H_2O(l) \qquad \Delta H = +6.02 \text{ kJ mol}^{-1} \qquad \Delta S_{sys} = +22 \text{ J K}^{-1} \text{ mol}^{-1}$$

The entropy change for the surroundings (ΔS_{surr}) depends on the transfer of heat to/from the surroundings, i.e. the enthalpy change (ΔH):

$$\Delta S_{surr} = -\frac{\Delta H}{T}$$

> Remember ΔH has to be in J mol^{-1} and the temperature in K.

Knowing the total entropy change enables a prediction to be made:

If ΔS_{total} is positive a reaction will occur spontaneously.

If $\Delta S_{total} = 0$ the reaction is at equilibrium.

> The total entropy change (ΔS_{total}) is the sum of these: $\Delta S_{total} = \Delta S_{sys} + \Delta S_{surr}$

Worked examples

1 Calculate the standard entropy change for the formation of ammonia, using the following standard entropies: $S^{\ominus}(N_2(g)) = 192 \text{ J K}^{-1}\text{mol}^{-1}$, $S^{\ominus}(H_2(g)) = 131 \text{ J K}^{-1}\text{mol}^{-1}$ and $S^{\ominus}(NH_3(g)) = 193 \text{ J K}^{-1}\text{mol}^{-1}$.

$$N_2(g) + 3H_2(g) \rightleftharpoons 2NH_3(g)$$

$$\Delta S^{\ominus} = \Sigma S^{\ominus} \text{ of products} - \Sigma S^{\ominus} \text{ of reactants} = 2 \times S^{\ominus}(NH_3(g)) - S^{\ominus}(N_2(g)) - 3 \times S^{\ominus}(H_2(g))$$

$$= 2 \times 193 - (192 + 3 \times 131) = -199 \text{ J K}^{-1} \text{mol}^{-1} S^{\ominus}$$

> There is a decrease in entropy because overall the number of moles of gas decreases.

2 Calculate the total entropy change for ice melting at 0°C.

$$\Delta S_{surr} = -\frac{\Delta H}{T} = -\frac{6020}{273} = -22 \text{ J K}^{-1} \text{mol}^{-1} \qquad \Delta S_{total} = \Delta S_{sys} + \Delta S_{surr} = +22 - 22 = 0$$

> Ice and water are equilibrium at 273 K.

? Quick check questions

1 Calculate the total entropy change for the decomposition of magnesium carbonate at 500°C. Does the reaction occur spontaneously?

$$MgCO_3(s) \rightarrow MgO(s) + CO_2(g) \qquad \Delta H = +100.3 \text{ kJ mol}^{-1}$$

$$\Delta S_{sys} = +174.8 \text{ J K}^{-1}\text{mol}^{-1}$$

2 What would be the minimum temperature required for any decomposition of magnesium carbonate?

Energy changes in solution

Chemical Ideas 4.5 and 3.2

When ionic compounds dissolve in water the ionic lattice is broken up, and the ions separate and become hydrated. Energy is used to break up the lattice (ionic bonds are broken) but energy is given out when the ions are hydrated (ion-dipole bonds are formed).

Lattice (formation) enthalpy, ΔH_{LE}, is the enthalpy change when 1 mole of solid is formed from its separate gaseous ions, e.g.

$$Na^+(g) + Cl^-(g) \rightarrow NaCl(s) \qquad \Delta H_{LE}(NaCl) = -788 \text{ kJ mol}^{-1}$$

$$Mg^{2+}(g) + 2Cl^-(g) \rightarrow MgCl_2(s) \qquad \Delta H_{LE}(MgCl_2) = -2434 \text{ kJ mol}^{-1}$$

This involves the **formation** of ionic bonds and so ΔH_{LE} is always negative. The bigger the lattice enthalpy, the stronger the ionic bonds. The strength of the ionic bonds depends on the charge and size of the ions.

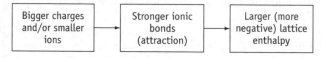

The ionic radius of elements depends on:

- nuclear charge (atomic number) – the bigger the nuclear charge the smaller the ion
- number of full energy levels – the more full levels the bigger the ion
- negative/positive charge – positive ions are smaller than negative ions with the same number of electrons.

Enthalpy of hydration, ΔH_{hyd}, is the enthalpy change when an aqueous solution is formed from 1 mole of gaseous ions, e.g.

$$Na^+(g) + aq \rightarrow Na^+(aq) \qquad \Delta H_{hyd}(Na^+) = -406 \text{ kJ mol}^{-1}$$

$$Br^-(g) + aq \rightarrow Br^-(aq) \qquad \Delta H_{hyd}(Br^-) = -337 \text{ kJ mol}^{-1}$$

> aq represents water acting as a solvent.

This involves ion-dipole attractions and so ΔH_{hyd} is always negative. The greater the enthalpy of hydration, the stronger the ion-dipole attractions and the greater the number of water molecules attached to the ion. The strength of the ion-dipole attraction depends on the charge and size of the ions.

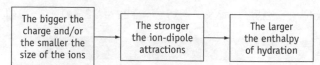

> Enthalpy of solvation, ΔH_{solv}, is the enthalpy change when a solution is formed, using a solvent other than water, from 1 mole of gaseous ions.

Enthalpy change of solution, $\Delta H_{solution}$, is the enthalpy change when 1 mole of solute dissolves to form a dilute solution. This can be thought of as a two-step process – breaking down the lattice and hydrating the gaseous ions produced – although it is easily measured experimentally:

$$\Delta H_{solution} = \Delta H_{hyd}(\text{cation}) + \Delta H_{hyd}(\text{anion}) - \Delta H_{LE}$$

> Breaking a solid lattice into gaseous ions is the reverse of lattice formation and so the energy input is $-\Delta H_{LE}$.

If $\Delta H_{solution}$ is negative, or slightly positive, then the solid will dissolve as the **entropy change** is generally favourable.

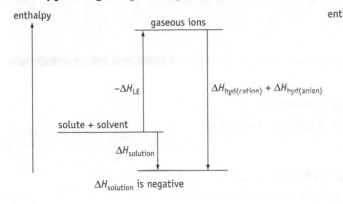

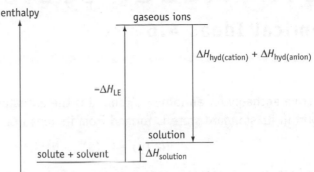

If $\Delta H_{solution}$ is large and positive then the solid will not dissolve, even though the **entropy change** is generally favourable, as too much energy is needed. This is always the case with non-polar solvents, as ΔH_{hyd} is tiny because there is little attraction between ion and solvent.

You may need to calculate $\Delta H_{solution}$ and predict the solubility of an ionic compound based on figures given to you.

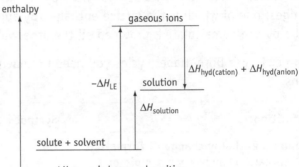

Worked example

Calculate $\Delta H_{solution}$ for $CaCl_2$ from the following data. Use this to predict if calcium chloride is water soluble.

$\Delta H_{hyd}(Ca^{2+}) = -1579$ kJ mol^{-1} $\Delta H_{hyd}(Cl^-) = -364$ kJ mol^{-1} $\Delta H_{LE}(CaCl_2) = -2255$ kJ mol^{-1}

$$\Delta H_{solution}(CaCl_2) = \Delta H_{hyd}(Ca^{2+}) + 2\Delta H_{hyd}(Cl^-) - \Delta H_{LE}(CaCl_2)$$

$$\Delta H_{solution}(CaCl_2) = (-1579) + (2 \times -364) - (-2255) = -52 \text{ kJ mol}^{-1}$$

Because $\Delta H_{solution}(CaCl_2)$ is negative, solid calcium chloride will dissolve in water.

▶ The value of $\Delta H_{hyd}(Cl^-)$ is multiplied by 2 because 2 moles of chloride ions are needed to produce 1 mole of calcium chloride.

? *Quick check questions*

1 Put the following ionic compounds in order from least to most negative ΔH_{LE}: LiF, KF, CaO and CaF$_2$.

2 Put the following ions in order from least to most negative ΔH_{hyd}: Al^{3+}, Na$^+$, Mg^{2+}, Ca^{2+} and K$^+$.

3 Draw enthalpy level diagrams for the dissolving of **(a)** KCl in water **(b)** CaCl$_2$ in water. ($\Delta H_{solution}(KCl) = +33$ kJ mol^{-1}, $\Delta H_{solution}(CaCl_2) = -52$ kJ mol^{-1})

4 Calculate $\Delta H_{solution}$ for AgI from the following data:

   $\Delta H_{hyd}(Ag^+) = -446$ kJ mol^{-1} $\Delta H_{hyd}(I^-) = -293$ kJ mol^{-1} $\Delta H_{LE} = -802$ kJ mol^{-1}

5 Would AgI be soluble in water? Give reasons for your answer.

Born–Haber cycles
Chemical Ideas 4.6

The lattice enthalpy for an ionic compound is the enthalpy change when 1 mole of the solid in its standard state is formed from its ions in the gaseous state, e.g.

$$Na^+(g) + Cl^-(g) \rightarrow NaCl(s)$$

Lattice enthalpies for ionic compounds are a good measure of the strength of ionic bonding in the lattice and are needed in any detailed discussion of solubility. They cannot be measured directly but can be calculated from experimental data using a Born–Haber cycle. This splits up the enthalpy of formation into individual enthalpy changes, one of which is the lattice enthalpy. By applying Hess's law the lattice enthalpy can be calculated, provided all the other values are known.

In order to use Born–Haber cycles you need to understand and learn the following terms:

> ▶ Remember to write in the correct state symbols for every equation you use.

> ▶ By definition, $\Delta H_f^{\ominus}$ of elements is taken to be zero.

Definition	Symbol	Example equations
Standard enthalpy change of formation. The enthalpy change when **1 mole** of a compound is **formed** from its elements, all in their standard states	$\Delta H_f^{\ominus}$	$Ca(s) + \frac{1}{2}O_2(g) \rightarrow CaO(s)$ $Na(s) + \frac{1}{2}Cl_2(g) \rightarrow NaCl(s)$
Standard enthalpy change of atomisation. The enthalpy change when **1 mole** of atoms in the gaseous state is **formed** from the element in its standard state	$\Delta H_{at}^{\ominus}$	$Na(s) \rightarrow Na(g)$ $\frac{1}{2}Cl_2(g) \rightarrow Cl(g)$
First ionisation enthalpy. The enthalpy change when 1 mole of +1 ions in their gaseous state are formed from 1 mole of gaseous atoms	$\Delta H_{i(1)}^{\ominus}$	$Na(g) \rightarrow Na^+(g)$ $Ca(g) \rightarrow Ca^+(g)$
Second ionisation enthalpy. The enthalpy change when 1 mole of +2 ions in their gaseous state is formed from 1 mole of gaseous +1 ions	$\Delta H_{i(2)}^{\ominus}$	$Na^+(g) \rightarrow Na^{2+}(g)$ $Ca^+(g) \rightarrow Ca^{2+}(g)$
First electron affinity. The enthalpy change when 1 mole of gaseous atoms each gain one electron	$\Delta H_{EA(1)}^{\ominus}$	$Cl(g) \rightarrow Cl^-(g)$ $O(g) \rightarrow O^-(g)$
Second electron affinity. The enthalpy change when 1 mole of gaseous −1 ions each gain one electron	$\Delta H_{EA(2)}^{\ominus}$	$Cl^-(g) \rightarrow Cl^{2-}(g)$ $O^-(g) \rightarrow O^{2-}(g)$
Lattice (formation) enthalpy. The enthalpy change when 1 mole of solid ionic compound is made from gaseous ions	$\Delta H_{LE}^{\ominus}$	$Na^+(g) + Cl^-(g) \rightarrow Na^+Cl^-(s)$

Born–Haber cycle

At each stage of the cycle, one change occurs, which is shown in bold.

Starting from the two elements in their standard states there are two routes round the Born–Haber cycle:

Route 1 – Going anticlockwise uses the enthalpy change of formation.

Route 2 – Going clockwise uses the standard enthalpy of atomisation (in order to obtain gaseous atoms) followed by ionisation or electron affinities (in order to obtain gaseous ions) and finally the lattice enthalpy.

Using Hess's Law:

enthalpy change of formation = enthalpies of atomisation + enthalpies of ionisation + electron affinities + lattice enthalpy

Rearranging this gives:

lattice enthalpy = enthalpy change of formation – enthalpies of atomisation – enthalpies of ionisation – electron affinities

Born–Haber cycle for sodium chloride

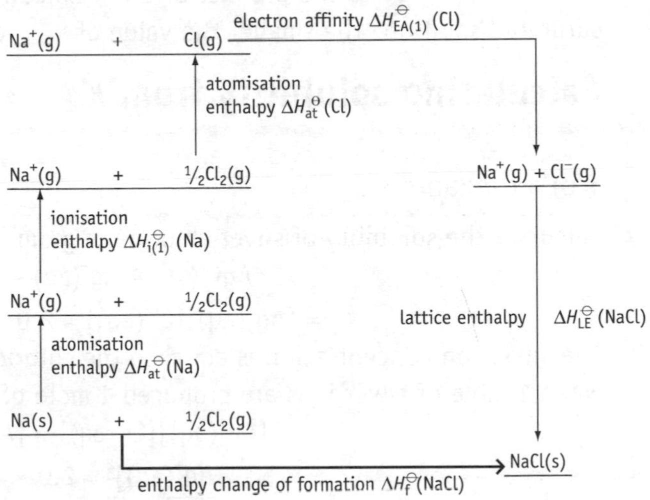

According to Hess's law:

ΔH_f(sodium chloride) = ΔH_{at}(sodium) + $\Delta H_{i(1)}$(sodium) + ΔH_{at}(chlorine) + $\Delta H_{EA(1)}$(chlorine) + ΔH_{LE}(sodium chloride)

Rearranging this gives:

ΔH_{LE}(sodium chloride) = ΔH_f(sodium chloride) – ΔH_{at}(sodium) – $\Delta H_{i(1)}$(sodium) – ΔH_{at}(chlorine) – $\Delta H_{EA(1)}$(chlorine)

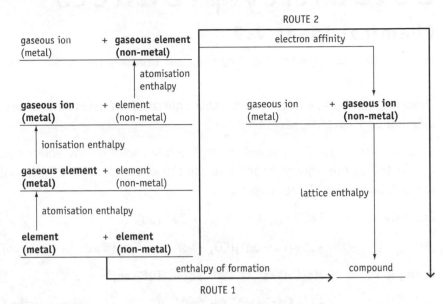

❓ Quick check questions

1 Calculate the lattice enthalpies of sodium and calcium chlorides, using the following values (all in kJ mol^{-1}):

ΔH_f(NaCl) = –411 ΔH_{at}(Na) = +107 $\Delta H_{i(1)}$(Na) = +502 $\Delta H_{EA(1)}$(Cl) = –355

ΔH_f(CaCl$_2$) = –796 ΔH_{at}(Ca) = +178 $\Delta H_{i(1)}$(Ca) = +596

ΔH_{at}(Cl) = +121 $\Delta H_{i(2)}$(Ca) = +1152

2 Account for the differences in the two values you obtained in question 1.

3 Draw a fully labelled Born–Haber cycle for potassium chloride.

$\Delta H_{at}^{\ominus}$ (K) = +89 kJ mol^{-1} $\Delta H_{i(1)}^{\ominus}$ (K) = +425 kJ mol^{-1} $\Delta H_f^{\ominus}$ (KCl) = –431 kJ mol^{-1}

> ► Remember, 1 mole of CaCl$_2$ needs 2 moles of chlorine atoms.

> ► Take care with the + and – signs before the numbers. Using brackets, for the sign and value, is the best idea.

Solubility products
Chemical Ideas 7.7

These are equilibrium constants for sparingly soluble (almost insoluble) ionic solids dissolving in water.

As long as some solid is present in the saturated solution the equilibrium will be unaffected by the amount of solid and so this does not appear in the solubility product expression. For example:

Example 1: $CaCO_3(s) \rightleftharpoons Ca^{2+}(aq) + CO_3^{2-}(aq)$

$K_{sp} = [Ca^{2+}(aq)][CO_3^{2-}(aq)]$ (K_{sp} units = mol^2dm^{-6})

Example 2: $Mg(OH)_2(s) \rightleftharpoons Mg^{2+}(aq) + 2OH^-(aq)$

$K_{sp} = [Mg^{2+}(aq)][OH^-(aq)]^2$ (K_{sp} units = mol^3dm^{-9})

The value of K_{sp} gives the product of the maximum concentrations possible in a saturated solution. The smaller the value of K_{sp} the less soluble is the solid.

> ▶ All insoluble solids are now treated as sparingly soluble, i.e. dissolve to a tiny extent.

> ▶ Remember that [ ] means concentration in $mol\ dm^{-3}$.

> ▶ K_{sp} is another example of an equilibrium constant.

Calculating solubility from K_{sp}

Worked example

Calculate the solubility of silver chloride in $g\ dm^{-3}$ ($K_{sp} = 2.0 \times 10^{-10}\ mol^2dm^{-6}$ at 298 K)

$$AgCl(s) \rightleftharpoons Ag^+(aq) + Cl^-(aq)$$

$$K_{sp} = [Ag^+(aq)][Cl^-(aq)] = 2.0 \times 10^{-10}\ mol^2dm^{-6}$$

The silver ion concentration is equal to the chloride ion concentration, because when 1 mole of silver ions are produced 1 mole of chloride ions are also produced:

$$[Ag^+(aq)][Cl^-(aq)] = [Ag^+(aq)]^2$$

$$[Ag^+(aq)]^2 = 2.0 \times 10^{-10}$$

$$[Ag^+(aq)] = \sqrt{[Ag^+(aq)]^2} = 1.4 \times 10^{-5}\ mol\ dm^{-3}$$

Solubility AgCl = $1.4 \times 10^{-5}\ mol\ dm^{-3}$ at 298 K

Converting the solubility to $g\ dm^{-3}$: Solubility AgCl = $1.4 \times 10^{-5} \times M_r$ (AgCl)

$$= 1.4 \times 10^{-5} \times 143.5 = 2.03 \times 10^{-3}\ g\ dm^{-3}\ at\ 298\ K$$

> ▶ Always quote the temperature because K_{sp} varies with temperature.

Predicting precipitation

If solutions containing the component ions are mixed and the product of the concentrations in the mixture exceeds K_{sp}, then a precipitate will form, until the product of the concentrations in solution agrees with K_{sp}.

Worked example

Equal volumes of $0.001\ mol\ dm^{-3}$ silver nitrate and $0.0001\ mol\ dm^{-3}$ sodium chloride solutions are mixed. Will a precipitate of silver chloride form? (K_{sp} AgCl = $2 \times 10^{-10}\ mol^2\ dm^{-6}$ at 298 K.)

Concentrations in the mixture are half the initial concentrations, because equal volumes are mixed. This means the same number of moles have been spread through twice the volume, so halving the concentrations.

$[Ag^+(aq)] = 0.0005$ mol dm^{-3} $[Cl^-(aq)] = 0.00005$ mol dm^{-3}
$[Ag^+(aq)][Cl^-(aq)] = 0.0005 \times 0.00005 = 2.5 \times 10^{-8}$ mol^{2}dm^{-6} at 298 K
This is greater than K_{sp} for silver chloride and so a precipitate forms.

Common ion effect

A sparingly soluble solid will be much less soluble in a solution containing one of the ions in the K_{sp} (i.e. common ion), than in water.

Worked example

What is the solubility of barium sulphate (a) in water (b) in 0.01 mol dm^{-3} sulphuric acid? (K_{sp} BaSO$_4$ = 1.0×10^{-10} mol^{2}dm^{-6} at 298 K)
$$BaSO_4(s) \rightleftharpoons Ba^{2+}(aq) + SO_4^{2-}(aq)$$
$$[Ba^{2+}(aq)][SO_4^{2-}(aq)] = 1.0 \times 10^{-10} \text{ mol}^2\text{dm}^{-6}$$

(a) The concentrations of barium and sulphate ions are equal, as seen from the equation:
$$BaSO_4(s) \rightleftharpoons Ba^{2+}(aq) + SO_4^{2-}(aq)$$
$$[Ba^{2+}(aq)]^2 = 1.0 \times 10^{-10} \text{ mol}^2\text{dm}^{-6}$$
$$[Ba^{2+}(aq)] = \sqrt{1.0 \times 10^{-10}} = 1.0 \times 10^{-5} \text{ mol dm}^{-3}$$
The solubility of BaSO$_4$ = 1.0×10^5 mol dm^{-3} at 298 K
Converting to g dm^{-3}: Solubility BaSO$_4$ = $1.0 \times 10^{-5} \times M_r$ BaSO$_4$
$$= 1.0 \times 10^{-5} \times 233 = 2.33 \times 10^{-3} \text{ g dm}^{-3} \text{ at 298 K}$$

(b) $[Ba^{2+}(aq)][SO_4^{2-}(aq)] = 1.0 \times 10^{-10}$ mol^{2}dm^{-6} at 298 K
Assuming all the sulphate ions are from the sulphuric acid, because the contribution from barium sulphate is negligible compared to this, $[SO_4^{2-}(aq)] = 0.01$ mol dm^{-3}
$$[Ba^{2+}(aq)] \times 0.01 = 1.0 \times 10^{-10} \text{ mol dm}^{-3} \text{ at 298 K}$$

$$[Ba^{2+}(aq)] = \frac{1.0 \times 10^{-10}}{0.01} = 1.0 \times 10^{-8} \text{ mol dm}^{-3}$$

Solubility BaSO$_4$ = 1.0×10^{-8} mol dm^{-3} at 298 K
Converting to g dm^{-3}:
Solubility BaSO$_4$ = $1.0 \times 10^{-8} \times M_r$ BaSO$_4$
$$= 1.0 \times 10^{-8} \times 233 = 2.33 \times 10^{-6} \text{ g dm}^{-3} \text{ at 298 K}$$
Barium sulphate is a thousand times less soluble in 0.01 M sulphuric acid than in water.

? Quick check questions

1 Explain what is meant by a saturated solution.

2 What is the common ion effect?

3 Write an expression for the solubility product for lead chloride (PbCl$_2$).

4 Calculate the solubility for silver iodide in g dm^{-3} (K_{sp} = 8.0×10^{-17} mol^2 dm^{-6}, A_r Ag = 107.9, A_r I = 126.9).

5 25 cm^3 of calcium chloride solution (0.0002 mol dm^{-3}) were added to 25 cm^3 of sodium carbonate solution (0.001 mol dm^{-3}). Does a precipitate form? Write an equation for the precipitation. (K_{sp} CaCO$_3$ = 5.0×10^{-9} mol^2 dm^{-6})

Weak acids and pH
Chemical Ideas 8.2

An acid is a proton (H^+) donor. In aqueous solution the acid donates protons to water molecules to form oxonium ions (H_3O^+). These are often abbreviated to $H^+(aq)$.

Acids vary in strength, i.e. in their ability to donate protons. In solutions of strong acids almost all of the acid molecules donate their protons. The acid is said to have undergone complete dissociation. Examples of strong acids are hydrochloric (HCl), sulphuric (H_2SO_4) and nitric (HNO_3) acids.

$$HCl(aq) \rightarrow H^+(aq) + Cl^-(aq)$$

$$H_2SO_4(aq) \rightarrow 2H^+(aq) + SO_4^{2-}(aq)$$

In weak acids only a small proportion of the acid molecules donate their protons. The acid is said to have undergone incomplete dissociation. The strength of an acid depends on the position of the equilibrium. The more the equilibrium lies to the right-hand side the stronger the acid. Examples of weak acids are carboxylic acids (including ethanoic acid) and carbonic acids.

$$CH_3COOH(aq) \quad \rightleftharpoons \quad CH_3COO^-(aq) \quad + \quad H^+(aq)$$

$$\text{ethanoic acid} \qquad\qquad \text{ethanoate}$$
$$\text{molecules} \qquad\qquad\quad \text{ions}$$

$$H_2CO_3(aq) \quad \rightleftharpoons \quad H^+(aq) + HCO_3^-(aq) \quad \rightleftharpoons \quad 2H^+(aq) + CO_3^{2-}(aq)$$

A good measure of the strength of a weak acid is the value of the equilibrium constant (K_a), called the acidity constant, or more often, the acid dissociation constant. The greater the value of K_a the stronger the acid.

For ethanoic acid:

$$K_a = \frac{[H^+(aq)][CH_3COO^-(aq)]}{[CH_3COOH(aq)]}$$

When comparing weak acids, which can have very small values, the K_a can be converted into a pK_a for ease of use:

$$pK_a = -\log K_a$$

Calculating pH

The pH scale is logarithmic and conveniently gives more manageable numbers for the acidity of solutions. For dilute solutions the normal range is 0–14.

$$pH = -\log [H^+(aq)]$$

Strong acids

Because a strong acid is fully dissociated we can assume that the concentration of acid put into the solution is the same as the concentration of the hydrogen ions in that solution [$H^+(aq)$], if the acid is monoprotic.

▶ Remember H^+ is a proton. You can use the terms interchangeably.

▶ Sometimes the word 'ionisation' is used instead of dissociation when referring to acids.

▶ Square brackets around a formula means the concentration of whatever is inside the brackets.

▶ Monoprotic acids have one acidic hydrogen.

Example 1 Calculating the pH of 0.001 mol dm^{-3} hydrochloric acid

$$HCl(aq) \rightarrow H^+(aq) + Cl^-(aq)$$

$$[H^+] = 0.001 \text{ mol dm}^{-3} \qquad pH = -\log 0.001 = 3$$

Example 2 Calculating the pH of 0.005 mol dm^{-3} sulphuric acid

$$\begin{array}{ccc} H_2SO_4(aq) & \rightarrow & 2H^+(aq) + SO_4^{2-}(aq) \\ 0.005 \text{ mol dm}^{-3} & & 0.01 \text{ mol dm}^{-3} \quad pH = -\log 0.01 = 2 \end{array}$$

> ◖ In this case the concentration of H⁺ is twice that of the H_2SO_4 because 1 mole of the acid dissociates to produce 2 moles of hydrogen ions.

Weak acids

Because a weak acid is only partially dissociated both the concentration of the acid and the dissociation constant are required in order to calculate the pH.

For example, calculating the pH of 0.01 M CH_3COOH ($K_a = 1.7 \times 10^{-5}$ mol dm^{-3} at 298 K):

$$CH_3COOH(aq) \rightleftharpoons CH_3COO^-(aq) + H^+(aq)$$

Assume $[CH_3COOH(aq)] = 0.01$ mol dm^{-3} because very few molecules dissociate.

> ◖ Learn these assumptions.

Assume $[H^+(aq)] = [CH_3COO^-(aq)]$ because for every mole of H⁺ produced there is 1 mole of CH_3COO^- produced. (Although a few protons are provided by the water this is insignificant compared to the protons provided by the acid, so is discounted.)

Putting these into the K_a expression:

$$1.7 \times 10^{-5} = \frac{[H^+(aq)]^2}{0.01}$$

$$[H^+(aq)] = \sqrt{1.7 \times 10^{-5} \times 0.01} = 4.1 \times 10^{-4} \text{ mol dm}^{-3}$$

$$pH = -\log(4.1 \times 10^{-4}) = 3.38 \text{ (at 298 K)}$$

> ◖ The temperature always needs to be quoted because pH varies with temperature.

Strong bases

Because a strong base is fully dissociated the concentration of base gives $[OH^-(aq)]$ directly.

For example, calculating the pH of 0.01 mol dm^{-3} NaOH:

$$\begin{array}{ccc} NaOH(aq) & \rightleftharpoons & Na^+(aq) + OH^-(aq) \\ 0.01 \text{ mol dm}^{-3} & & 0.01 \text{ mol dm}^{-3} \end{array}$$

To calculate $[H^+(aq)]$, using the already known $[OH^-(aq)]$, another expression is used, the ionic product for water. Water dissociates slightly:

$$H_2O(l) \rightleftharpoons [H^+(aq)] + [OH^-(aq)]$$

$$K_w = [H^+(aq)][OH^-(aq)] = 10^{-14} \text{ mol}^2 \text{ dm}^{-6} \text{(at 298 K)}$$

$$10^{-14} = [H^+(aq)] \times 0.01 \qquad [H^+(aq)] = 10^{-12} \text{ mol dm}^{-3}$$

$$pH = 12$$

> ◖ K_w is the symbol used to represent the ionic product for water.

? *Quick check questions*

1 Calculate the pH of 0.001 mol dm^{-3} nitric acid.

2 Calculate the pH of 0.05 mol dm^{-3} potassium hydroxide.

3 Give the expression for the K_a of methanoic acid (HCOOH).

4 Calculate the pH of 0.001 mol dm^{-3} methanoic acid ($K_a = 1.6 \times 10^{-4}$ mol dm^{-3}, p$K_a = 3.8$).

Buffer solutions

Chemical Ideas 8.3

These are solutions that have a constant pH, despite dilution or small additions of acid or alkali. Buffer solutions contain either a weak acid and its salt, e.g. ethanoic acid and sodium ethanoate, or a weak base and its salt, e.g. ammonia and ammonium chloride.

All buffer solutions contain large amounts of a proton donor, i.e. weak acid or conjugate acid, and large amounts of a proton acceptor, i.e. weak base or conjugate base. Any additions of acid or alkali react with these large amounts (buffers) and so this keeps the pH constant.

For example, ethanoic acid and sodium ethanoate solution. The weak acid (ethanoic) partially dissociates to produce its conjugate base and protons. In the case of a buffer solution the acid used is a weak acid so the position of equilibrium lies to the left in the equation below:

$$CH_3COOH(aq) \rightleftharpoons CH_3COO^-(aq) + H^+(aq) \quad \text{(equation 1)}$$
$$\text{weak acid} \qquad\qquad \text{conjugate base}$$

The salt (sodium ethanoate) dissolves almost completely in water and the position of equilibrium lies to the right in the equation below:

$$H_3COONa(aq) \rightarrow CH_3COO^-(aq) + Na^+(aq) \quad \text{(equation 2)}$$
$$\text{sodium ethanoate} \qquad \text{ethanoate ion}$$

If small amounts of acid are added the H^+ can be removed by reaction with the anion present in large amounts from the salt, i.e. the position of equilibrium moves to the left in equation 2, so maintaining the pH.

If small amounts of alkali are added the weak acid dissociates to produce more H^+, which react with the extra hydroxide ions, i.e. the position of equilibrium moves to the right in equation 1, so maintaining the pH.

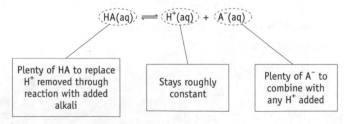

$$HA(aq) \rightleftharpoons H^+(aq) + A^-(aq)$$

| Plenty of HA to replace H^+ removed through reaction with added alkali | Stays roughly constant | Plenty of A^- to combine with any H^+ added |

Calculations with buffers

These all start from the K_a expression:

$$K_a = \frac{[H^+(aq)][A^-(aq)]}{[HA(aq)]}$$

Using the assumptions on the right, $K_a = \dfrac{[H^+(aq)][\text{salt}]}{[\text{acid}]}$

Two assumptions have been made here, which you need to know about. Firstly, assume all the anions (A^-) have come from the salt, the contribution from the acid is negligible. Secondly, the concentration of the acid in solution [HA(aq)] is the same as the amount of acid put into the solution (i.e. discount any dissociation).

Finding the pH of a buffer solution

If the K_a is known, together with the concentrations of salt and weak acid, then the hydrogen ion concentration can be calculated and hence the pH.

Worked example

Calculate the pH of a buffer made by mixing equal quantities of 0.1 mol dm^{-3} ethanoic acid and 0.1 mol dm^{-3} sodium ethanoate solutions (for ethanoic acid $K_a = 1.7 \times 10^{-5}$ mol dm^{-3} at 298 K).

By mixing equal quantities, each original concentration will be halved, so:

$$[CH_3COOH(aq)] = 0.05 \text{ mol dm}^{-3} \quad [CH_3COO^-(aq)] = 0.05 \text{ mol dm}^{-3}$$

$$1.7 \times 10^{-5} = \frac{[H^+(aq)] \times 0.05}{0.05} \quad pH = -\log[H^+(aq)] = 4.77$$

Making a buffer of a specified pH

Using the general expression:

$$K_a = \frac{[H^+(aq)][salt]}{[weak\ acid]}$$

The value of K_a determines the pH range for the buffer, whereas the ratio [salt] : [weak acid] determines the exact pH in this range. So to make a buffer the most important factor is to choose the weak acid with the correct K_a (or pK_a) and then calculate the ratio of [salt] : [weak acid].

$pK_a = -\log K_a$

? Quick check questions

1 Explain what a buffer solution is and how it works.

2 Calculate the pH of a solution containing 0.001 M ethanoic acid and 0.005 M sodium ethanoate.

3 Which weak acid/salt would you use to make a buffer of pH 3.1 and what would their concentrations be in the mixture?

Acid	K_a mol dm^{-3}	pK_a
methanoic	1.6×10^{-4}	3.8
ethanoic	1.7×10^{-5}	4.8
propanoic	1.3×10^{-5}	4.9
benzoic	6.3×10^{-5}	4.2
chloroethanoic	1.2×10^{-3}	2.9

[Hint: if the concentrations of acid and salt are the same then $K_a = [H^+]$, so pK_a = pH. This allows you to decide which acid/salt system to use.

Medicines by Design (MD)

New medicines have made a major impact on all our lives. This unit is about the development and synthesis of more effective drugs. CI refers to sections in your Chemical Ideas textbook.

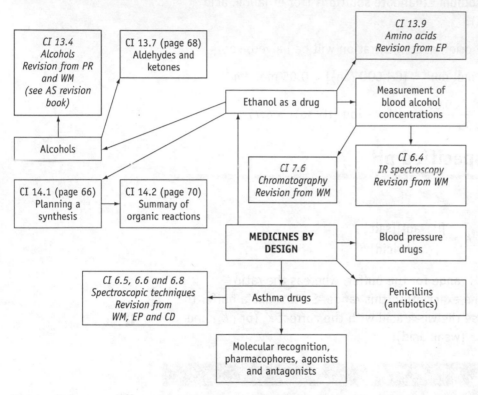

In this unit the study of these Chemical Ideas runs parallel with all the MD Storylines.

This unit is very different in structure in that it is predominately **synoptic** in nature, i.e. it pulls together many of the ideas and concepts you have come across earlier in the course. The main thrust of this unit is to pull together all the organic chemistry from the two years of the course as well as spectroscopic interpretation and use this to synthesise and analyse compounds.

You will need to use the organic reactions in the following sections in order to synthesise new compounds. You need to learn the reaction conditions for these reactions and to be able to identify the type of reaction.

You will also need to calculate **percentage yields**. For example, in an experiment 9.0 g of ethanol are oxidised using excess potassium dichromate solution under reflux. After distillation 10.1 g of ethanoic acid were recovered. Calculate the percentage yield.

$$\text{Moles of ethanol} = \text{mass (g)}/M_r = 9.0/46 = 0.196$$

Because 1 mole of ethanol yields 1 mole of ethanoic acid, if there is 100% yield we would expect the theoretical yield (mass) of ethanoic acid to be:

$$\text{Number of moles} \times M_r = 0.196 \times 60 = 11.7 \text{ g}$$

$$\text{Percentage yield} = \frac{\text{experimental yield}}{\text{theoretical yield}} = \frac{10.1}{11.7} \times 100 = 86\%$$

The **overall yield** of a series of reactions will determine which route is taken in an organic synthesis.

The action of drugs
Storyline MD

Just as enzymes have active sites, there are many **receptor** sites in biological systems, particularly at nerve endings.

These receptor sites have a specific three-dimensional shape. The natural neurotransmitter not only has the correct shape to fit into the receptor, but forms bonds at several points in order to produce a response. This process is called **molecular recognition**. Chemists try to identify the part of the natural molecule (**pharmacophore**) that produces this pharmacological activity (biological response), so that new drugs can be designed.

Computer graphics can be used to produce a 3D map of the receptor and of promising molecules. Molecules with both the correct shape and orientation of the functional groups should be most effective at producing a response. This shortens the time needed to design new drugs.

Agonists mimic the natural neurotransmitter by fitting, binding and producing a response. These molecules contain the pharmacophore.

Antagonists fit the receptor but do not bind at all the correct points and so produce no response. The molecule only contains part of the pharmacaphore.

For example:

- Natural neurotransmitters, such as adrenaline, often produce a response at several similar receptors leading to a range of changes, e.g. higher blood pressure, increased heart rate, widening of the airways in lungs, increased sweating. A new asthma drug has to bind and produce a response only at the receptors on the muscles of the airways, e.g. salbutamol (agonist).

- Beta blockers are antagonists for the noradrenaline receptors on heart muscle and so reduce maximum heart rate.

- Ethanol depresses the activity of the central nervous system by binding near the receptor site and making the natural neurotransmitter more effective.

- Captopril inhibits one of the enzymes that is part of the natural process for increasing blood pressure and so is used to treat patients with high blood pressure.

- Antibiotics kill (disease causing) bacteria. Penicillin works by inhibiting a bacterial enzyme, whose function is to strengthen cell walls by forming crosslinks. All new bacteria thus have a weakened cell wall.

> Make sure you know the meaning of all the terms in bold type.

Aldehydes and ketones
Chemical Ideas 13.7

These two homologous series both contain the **carbonyl group**: $\ce{C=O}$

This functional group gives the same chemical properties to both aldehydes and ketones. However, the hydrogen atom attached to the aldehydic carbonyl group gives some distinctive properties that enable the two types of compound to be distinguished. The names and chemical structures of some common aldehydes and ketones are given below.

ethanal

CH_3CH_2CHO
propanal

propanone

$CH_3COCH_2CH_3$
butanone

> The names of aldehydes end with -al and ketones with -one. Remember that the carbon of the carbonyl counts as one of the carbons in the chain.

Preparation

Both aldehydes and ketones are easily prepared by oxidation of the correct alcohol. For an aldehyde, the reactant is a primary alcohol. Mild oxidation is needed and the aldehyde is distilled out as it is formed to prevent further oxidation to the acid, e.g.

$$CH_3CH_2OH \xrightarrow[\text{reflux}]{H^+/Cr_2O_7^{2-}} CH_3CHO + H_2O$$
ethanol → ethanal

For a ketone, the reactant is a secondary alcohol, e.g.

$$CH_3CHOHCH_3 \xrightarrow[\text{reflux}]{H^+/Cr_2O_7^{2-}} CH_3COCH_3 + H_2O$$
propan-2-ol → propanone

In both cases the laboratory reagent is acidified potassium dichromate solution, which changes from orange to green as the reaction proceeds.

The dichromate is reduced to Cr^{3+} in the reaction

$$Cr_2O_7^{2-}(aq) + 14H^+(aq) + 6e^- \rightarrow 2Cr^{3+}(aq) + 7H_2O$$
orange → green

> Acidified potassium dichromate solution is made by dissolving potassium dichromate into dilute sulphuric acid and the formula is often abbreviated to $H^+/Cr_2O_7^{2-}$.

Reactions

Oxidation

Only aldehydes can be further oxidised with acidified dichromate because of the hydrogen atom on the carbonyl group, e.g.

$$CH_3CHO \xrightarrow[\text{heat}]{H^+/Cr_2O_7^{2-}} CH_3COOH$$
ethanal → ethanoic acid

> Warm acidified dichromate can be used to distinquish between an aldehyde and a ketone. Only the aldehyde causes a colour change from orange to green. But remember primary and secondary alcohols also give this colour change with the same conditions.

Reduction

The powerful reducing agent, sodium tetrahydridoborate ($NaBH_4$), reduces carbonyl compounds back to the alcohol, e.g.

$$CH_3CHO \xrightarrow{NaBH_4} CH_3CH_2OH$$
ethanal ethanol

$$CH_3COCH_3 \xrightarrow{NaBH_4} CH_3CHOHCH_3$$
propanone propan-2-ol

Addition

All aldehydes and ketones undergo nucleophilic addition reactions because the carbon of the carbonyl group has a slight positive charge.

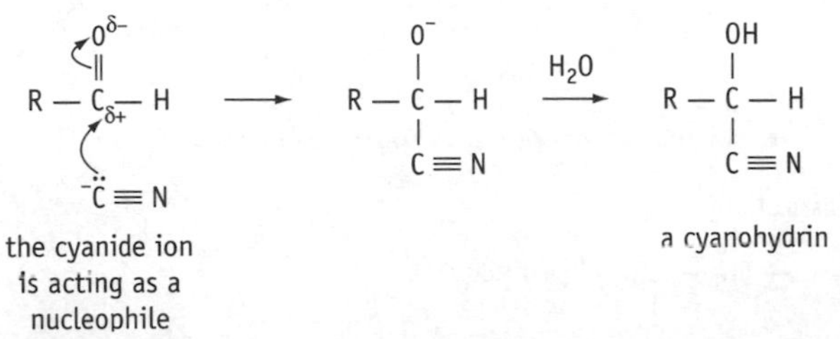

the cyanide ion is acting as a nucleophile

a cyanohydrin

This type of reaction is useful because it produces a new C—C bond adding another carbon into a molecule.

? Quick check questions

1 Give the formula and name for the alcohol that is oxidised to produce:

   (a) butanone (b) pentanal (c) propanoic acid.

2 What would you see if (a) butanal and (b) butanone were each separately mixed with warm acidified potassium dichromate?

A summary of organic reactions

Chemical Ideas 14.2

For each of the following reactions be sure you know the reaction conditions and the reaction type. You will also have to be able to write balanced equations for these reactions.

Alkenes

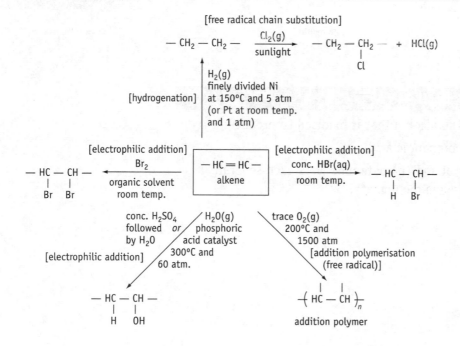

[free radical chain substitution]

$$- CH_2 - CH_2 - \xrightarrow[\text{sunlight}]{Cl_2(g)} \quad - CH_2 - CH_2 - \quad + \quad HCl(g)$$
                                                    $|$
                                                    Cl

[hydrogenation] $\quad$ $H_2(g)$
finely divided Ni
at 150°C and 5 atm
(or Pt at room temp.
and 1 atm)

[electrophilic addition] $\quad$ $\xleftarrow[\substack{\text{organic solvent} \\ \text{room temp.}}]{Br_2}$ $\quad$ $-HC = HC-$ $\quad$ $\xrightarrow[\text{room temp.}]{\text{conc. HBr(aq)}}$

$-HC - CH-$
$\quad |\quad\;\; |$
$\quad Br\;\; Br$

alkene

$-HC - CH-$
$\quad |\quad\; |$
$\quad H\;\; Br$

[electrophilic addition]

conc. H_2SO_4 $\quad$ $H_2O(g)$
followed $\quad or$ $\quad$ phosphoric
by H_2O $\quad$ acid catalyst
$\quad\quad\quad\quad$ 300°C and
$\quad\quad\quad\quad$ 60 atm.

trace $O_2(g)$
200°C and
1500 atm
[addition polymerisation
(free radical)]

$-HC - CH-$
$\quad |\quad\; |$
$\quad H\;\; OH$

$\left(\!\!\begin{array}{c} | \quad\; | \\ HC - CH \\ \end{array}\!\!\right)_n$

addition polymer

Halogenoalkanes

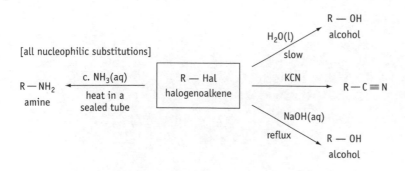

[all nucleophilic substitutions]

$\quad\quad\quad\quad\quad\quad\quad\quad\quad$ R — OH
$\quad\quad\quad\quad\quad\quad\quad\quad\quad$ alcohol

$\quad\quad\quad\quad\quad$ $H_2O(l)$
$\quad\quad\quad\quad\quad$ slow

$R - NH_2$ $\xleftarrow[\substack{\text{heat in a} \\ \text{sealed tube}}]{\text{c. } NH_3(aq)}$ $\boxed{\substack{R - Hal \\ \text{halogenoalkene}}}$ $\xrightarrow{\text{KCN}}$ $R - C \equiv N$

amine

$\quad\quad\quad\quad\quad$ NaOH(aq)
$\quad\quad\quad\quad\quad$ reflux $\quad$ R — OH
$\quad\quad\quad\quad\quad\quad\quad\quad\quad$ alcohol

Alcohols

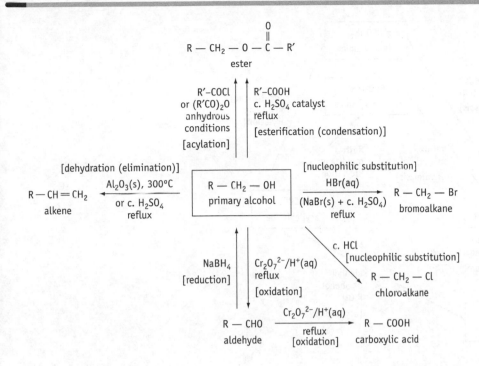

R — CH₂ — O — C — R' — ester

R'–COCl or (R'CO)₂O anhydrous conditions [acylation]

R'–COOH c. H₂SO₄ catalyst reflux [esterification (condensation)]

[dehydration (elimination)]

R — CH = CH₂ alkene

Al₂O₃(s), 300°C or c. H₂SO₄ reflux

R — CH₂ — OH primary alcohol

[nucleophilic substitution]

HBr(aq) (NaBr(s) + c. H₂SO₄) reflux

R — CH₂ — Br bromoalkane

c. HCl [nucleophilic substitution]

R — CH₂ — Cl chloroalkane

NaBH₄ [reduction]

Cr₂O₇²⁻/H⁺(aq) reflux [oxidation]

R — CHO aldehyde

Cr₂O₇²⁻/H⁺(aq) reflux [oxidation]

R — COOH carboxylic acid

Remember a secondary alcohol will oxidise to a ketone under reflux with acidified potassium dichromate while a tertiary alcohol will not oxidise.

Amines

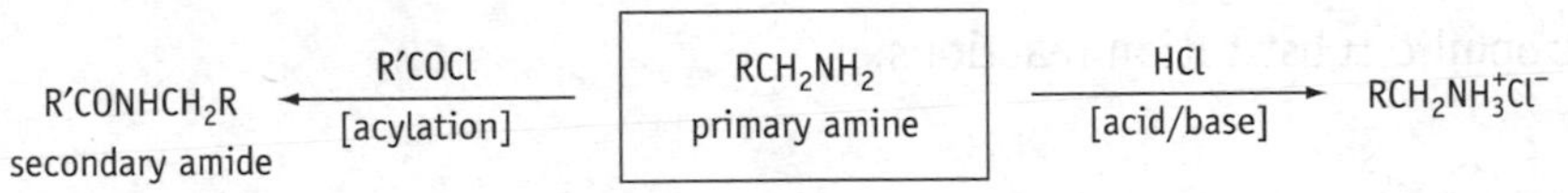

R'CONHCH₂R secondary amide

R'COCl [acylation]

RCH₂NH₂ primary amine

HCl [acid/base]

RCH₂NH₃⁺Cl⁻

Aldehydes

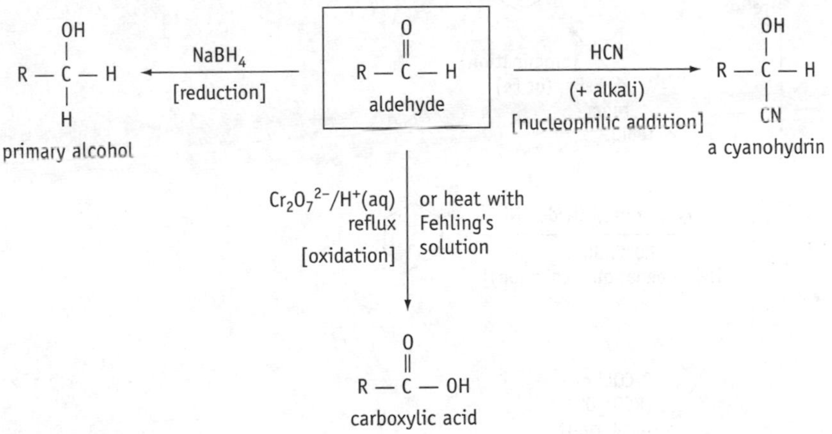

OH
R — C — H
H
primary alcohol

NaBH₄ [reduction]

O
R — C — H
aldehyde

HCN (+ alkali) [nucleophilic addition]

OH
R — C — H
CN
a cyanohydrin

Cr₂O₇²⁻/H⁺(aq) reflux [oxidation]

or heat with Fehling's solution

O
R — C — OH
carboxylic acid

Ketones

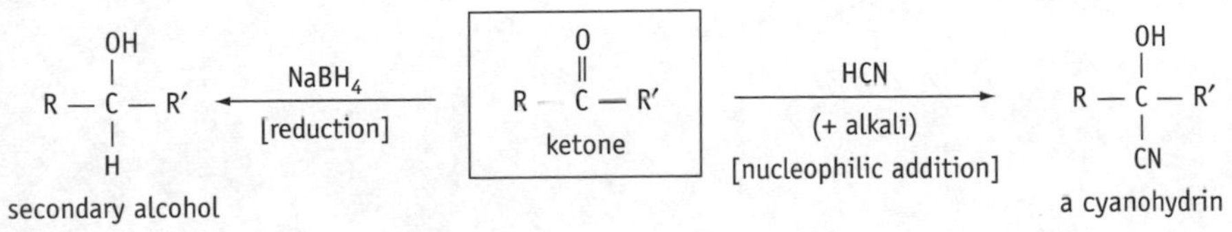

OH
R — C — R'
H
secondary alcohol

NaBH₄ [reduction]

O
R — C — R'
ketone

HCN (+ alkali) [nucleophilic addition]

OH
R — C — R'
CN
a cyanohydrin

Carboxylic acid and derivatives

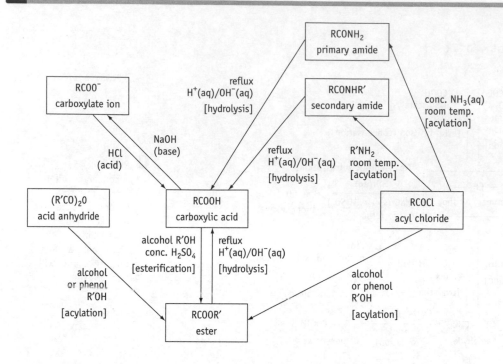

Arenes

All the reactions in this section are electrophilic substitution reactions.

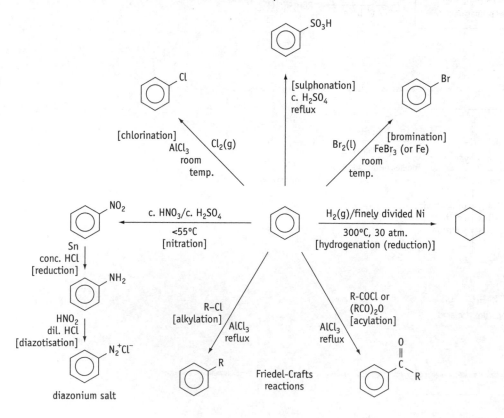

Quick check questions

1 Draw a full structural formula for the following and name the functional groups:

   (a) $CH_3CH_2CONHC_6H_5$ (b) $C_6H_5COOCH_3$ (c) CH_3COCl

2 Give equations for the hydrolysis of each compound in question 1, under acid conditions.

3 Methane can be converted into methanol by a two-step process.

   (a) Name the intermediate.

   (b) Give the reagents and conditions for each step.

   (c) Write equations for each step.

4 Methylbenzene can be converted into 2-methylphenylamine by a two-step process.

   (a) Name the intermediate.

   (b) Give the reagents and conditions for each step.

   (c) Write equations for each step.

5 Give equations for each of the following, stating conditions:

   (a) Hydrogenation of cyclohexene.

   (b) Neutralisation of propylamine with hydrochloric acid.

   (c) Dehydration of propan-2-ol.

Visiting the Chemical Industry (VCI)

Chemical Ideas 15.1–15.5

- Questions relating to the chemical industry may be examined in any of the written exam papers.

- Questions specifically relating to Visiting the Chemical Industry (VCI) may be asked in 2854 Chemistry by Design.

Before starting a full-scale plant production a process is trialled on a small scale, usually laboratory scale (gram scale) then pilot plant (kilogram scale) and finally full plant scale.

Most chemical processes involve the sequence – feedstock preparation, reaction and separation – as in the following diagram.

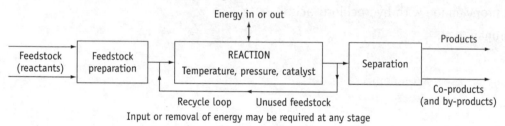

Any production process can be **batch** or **continuous** and must be carried out as safely as possible.

	Batch	Continuous
Description	Stages occur at separate times in sequence. After the reaction is over the vessel is emptied and the process started again.	Reactants are entering the reaction vessel, the reaction is taking place and products are being removed continuously.
Examples	Dyes and steels Pharmaceuticals such as paracetamol	Iron Sulphuric acid
Advantages	Cost effective for small quantities Range of products can be made in one vessel	More easily automated Lower numbers of process operators needed
Disadvantages	Risk of cross-contamination increased Larger workforce required	High capital costs to build plant Not cost effective if running below capacity

The main **construction material** for the plant is mild steel for economy, but where corrosion resistance is required then more expensive materials are used, e.g. glass-lined steel, stainless steel or reinforced plastics.

Feedstocks are the reactants needed for the chemical process. These are prepared from **raw materials** present in the natural environment, e.g. natural gas, crude oil, coal, limestone, etc.

The **optimum conditions** (temperature, pressure and catalyst) for a chemical process are determined by laboratory research. The aim is to maximise throughput

> Don't panic if you don't recognise the process used in a Visiting the Chemical Industry type question. If this is the case there will be enough information given in the question to use the ideas here.

> For a more extensive list of advantages and disadvantages see page 356 of Chemical Ideas.

(i.e. maximum amount of product in a minimum time). This means that the lowest temperature/pressure and cheapest catalyst (consistent with this aim) will be chosen. The conditions are often a compromise between rate and yield, particularly for reversible reactions, which cannot go to completion but attain equilibrium.

Co-products are the products other than the desired one formed in a reaction, e.g. in the equation below phenol is the desired product but for the reaction to occur, propanone is also produced as a co-product:

$$C_6H_5H(CH_3) + O_2 \rightarrow C_6H_5OH + (CH_3)_2CO$$
$$\text{cumene} \qquad\qquad \text{phenol} \qquad \text{propanone}$$

Co-products are unavoidable. If they can be sold it increases plant profitability.

By-products are produced as a result of unwanted side reactions and do not appear in the reactor equation. Conditions can usually be chosen to minimise these side reactions and so by-products are produced in much smaller quantities than products or co-products.

The separation stage produces the pure product, unreacted feedstock, any co-products and by-products. The unreacted feedstock is **recycled** to the reactor and increases the efficiency of the process. This also minimises waste effluent.

Energy is expensive and so the chemical engineer designs the plant with **energy efficiency** in mind. Exothermic reactions give out heat which, by using a heat exchanger, can produce hot water or steam that can be used in other processes.

Some major **costs**, e.g. research and development, plant design, construction and start-up are incurred before the plant starts production.

Sales of product have to cover these costs and the cost of production before a profit can be made. Production costs are of two types – fixed or variable:

- **Fixed** costs have to be paid no matter how much product is produced. The cost of construction is one of these (often referred to as a **capital** cost). Labour costs are another fixed cost.

- **Variable** costs depend on the level of production. If there is no production then there are no variable costs. The greater the production the greater the variable costs are. Some examples are raw materials, effluent treatment and distribution.

> You will need to apply the terms in bold to any chemical process. It is a good idea to learn the definitions of these words.

? Quick check questions

1 Explain what a batch process is.

2 Explain why batch processes are preferred for the production of azo dyes.

3 **(a)** What are feedstocks?

   (b) Name the feedstocks used for the manufacture of ammonia in the Haber process.

   (c) Name the raw materials used for the manufacture of ammonia in the Haber process.

4 Locating a new chemical plant on an existing site of chemical manufacture leads to lower costs. Why?

Experimental techniques

Make sure you are familiar with the following experimental techniques, as well as those used in the AS year.

1. Using a colorimeter

a Select a filter with the complementary colour to the solution being tested.

b Zero the colorimeter using a tube/cuvette of distilled water.

c If appropriate, measure the absorbance of a series of solutions of known concentration. Plot a calibration curve (absorbance vs concentration).

d Measure the absorbance of the unknown solution. Read off its concentration on the calibration curve.

> This technique is used to find the concentration of a coloured solution.

> **Example:** Finding the concentration of manganate(VIII) ions (Activity SS1.1).

2. Measuring a melting point

a Introduce a small amount of dry solid into a melting-point tube; tap the tube so that the solid falls to the bottom (sealed end).

b Place the tube in the melting-point apparatus and heat gently.

c Note the temperature at which the solid melts and the melting range.

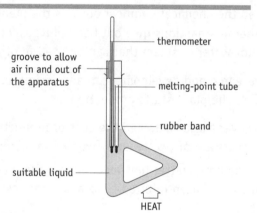

groove to allow air in and out of the apparatus — thermometer — melting-point tube — rubber band — suitable liquid — HEAT

> A compound can be identified by its melting point. Pure compounds have sharp melting points.

> **Example:** Testing the purity of hexanedioic acid (Activity DP2.2).

3. Thin layer chromatography (t.l.c.)

a Spot any mixtures and reference samples on a pencil line 1 cm from the base of a t.l.c. plate.

b Place the t.l.c. plate in a beaker containing solvent (making sure the solvent is below the pencil line). Cover the beaker with a watch glass.

c Remove the t.l.c. plate when the solvent is near the top. Mark how far the solvent had reached.

d Locate any spots with iodine, ninhydrin or under an ultraviolet lamp.

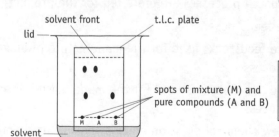

lid — solvent front — t.l.c. plate — spots of mixture (M) and pure compounds (A and B) — solvent — M A B

> This technique allows the identification of components in a mixture.

> **Example:** Preparing salicylic acid (Activity WM2).

4. Refluxing

a Add a few anti-bumping granules to the reagents. Do NOT stopper.

b Connect the condenser to the water supply.

c Heat so the liquid boils gently, using a Bunsen or heating mantle.

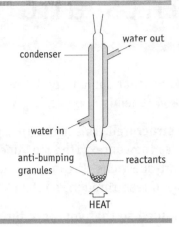

> Refluxing ensures volatile reactants or products don't escape while the reaction is in progress.

Example: Breaking down nylon-6,6 (Activity DP2.2).

5. Recrystallisation

a Dissolve the solid in a minimum quantity of hot solvent.

b Filter, and retain the filtrate.

c Allow the filtrate to cool until crystals form.

d Collect the crystals by vacuum filtration. Dry in air or a desiccator.

> This technique is used to purify an organic solid.

Example: Purification of aspirin (Activity WM5.1).

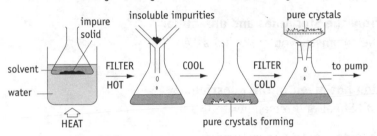

6. Measuring a cell emf

a Construct the half-cell whose electrode potential you wish to measure. Ensure all solutions have concentration 1.0 mol dm^{-3} and are at 25°C.

b Connect the half-cell to a standard hydrogen half-cell or other reference cell (see diagram) using a high-resistance voltmeter and salt bridge.

c Check the reading on the voltmeter is positive – if it is then the half-cell connected to the positive terminal of the voltmeter is the positive electrode.

d Record the voltmeter reading – this is the cell emf (E_{cell}).

e Standard electrode potentials can be measured by connecting any half-cell to a standard hydrogen half-cell.

> This technique can also be used to find standard electrode potentials.

Example: Investigating electrochemical cells (Activity SS3.2).

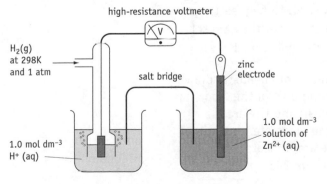

Exam hints and tips

Here are some general points that apply to both Module 2849 (Chemistry of Materials) and Module 2854 (Chemistry by Design) written exam papers.

1 All the questions are structured. This means that you are given a 'stem' of information which provides the context (the Storyline) for the question. This is followed by a series of part-questions. It can be quite helpful to underline key pieces of information as you read through the stem.

2 The questions are designed so that you work through them in stages. Answers to the part-questions are often linked together and are linked to the information you are given. There may be additional 'mini-stems' leading into a group of part-questions. So, remember to look back at both the information in the stem and your earlier answers. Work through the questions in order – don't cherry pick.

3 Part-questions are linked by the context – not by chemical topic. You will need to dip into several different parts of your knowledge in one question.

4 Contexts will be a mixture of familiar ones from the Storylines and unfamiliar ones. Don't panic if the context is unfamiliar – the chemistry you are being asked will be familiar.

5 The Data Sheet provides additional information not given in the question. Make sure you are familiar with what is on the Data Sheet and remember to use it.

6 All questions are compulsory – you have no choice. Try to answer every question. It is better to make a sensible guess, which could score you one mark, than to put nothing at all. You cannot get negative marks. You have nothing to lose by guessing – so use your chemical common sense!

7 Examiners use an agreed mark scheme. The marks are given for very specific points, so you must be precise and use language accurately. However, the mark scheme is flexible and you will always get credit for something that is chemically sensible. Remember this is an A level chemistry exam, so write in A level chemistry language, not language for GCSE chemistry or A level biology.

8 Some questions will require knowledge of applications of chemistry and the work of chemists. This means that it is a good idea to have read and made notes on the Storylines, particularly those sections listed in the 'Check your Notes' activities at the end of each unit.

9 The first question is designed to be quite straightforward, to help settle you into the exam. So, you are advised to do the questions in the order they are set. Often, for some reason, candidates do their best work in later questions!

Unit 2854 is a synoptic paper. It will draw upon a wide range of chemistry drawn from right across the two years of the course. Each teaching unit in this book has an introductory page consisting of a spider diagram summarising the areas covered in that unit. In the units from 2854 some of these areas are highlighted in italics as areas previously covered in the course. These are indicators that this material may be incorporated into a synoptic question on the 2854 paper.

Some tips

- Give oxidation states with a sign, i.e. you must write + or − before the number (e.g. +3, −5).

- Enthalpy changes **must** have a sign as well as a unit (e.g. −541 kJ mol^{-1}). Remember: sign, number, unit.

- Give state symbols in chemical equations only when requested. They are often asked for in reactions where there is a change of state.

- Use chemical language correctly (e.g. make sure you know the difference between atom, molecule and ion, between chlorine and chloride, and between hydroxide and hydroxyl).

$C_3H_6O_2$ C_2H_5COOH

molecular formula   shortened structural formula   full structural formula   skeletal formula

- Know the difference between molecular, structural, full structural and skeletal formulae.

- Draw dot and cross diagrams properly, showing only outer shell electrons, and not forgetting lone pairs of electrons.

lone pair of electrons shown

Quality of written communication

Both written exams will contain marks for the quality of written communication (QWC) included in some sections of extended writing. It will be made clear which answers will be marked for QWC. To obtain these marks you just need to:

- use correct scientific language

- write in full sentences (not bullet points like this text)

- write a minimum of two full sentences, spelt correctly and using correct grammar.

Calculations

- Always show your working. Examiners use consequential marking where possible to give credit for correct working even if your final answer is wrong, so make sure you include everything.

- Write your final answer on the answer line.

- Always give units (unless already present) and a sign if appropriate (<u>always</u> for enthalpy changes).

- Give the answer to the same number of significant figures as in the data in the question (usually 2 or 3 significant figures).

- Always think about whether your answer is sensible (e.g. ΔH_c values are always negative).

Diagrams

- Make sure the lines on diagrams are clear and unambiguous, not 'sketchy'.

- Use a pencil and always label clearly.

- For experimental work, draw a cross-section not 3D.

- Check that any joints you have drawn look airtight and that there is an unobstructed path through the apparatus for any liquid or vapour. Think whether the apparatus <u>as a whole</u> should be airtight or not (e.g. heated equipment, such as apparatus for heating under reflux or carrying out a distillation, should be open to the air at one end).

- Use an arrow labelled 'heat' to represent a Bunsen.

For more details of the experimental techniques you need to know about see pages 76–77.

Practice exam questions

Module 2849

1 PET, polyethylene terephthalate, is a polymer that has been used as a fibre for many years. More recently, plastics made from PET have replaced glass as the preferred container for fizzy drinks. Like glass, plastics made from PET are transparent and gases do not diffuse through them to any appreciable extent. This is especially important in food packaging.

> ► Think of everyday uses.

  a Suggest a reason why it is important for plastics made from PET to be airtight. [1]

The repeating unit of PET is shown below.

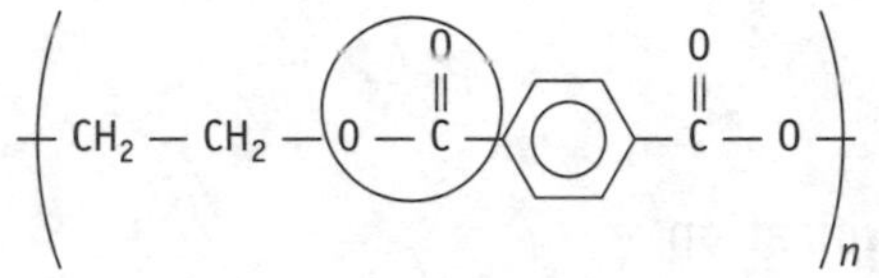

> ► Make sure you give the name and not the structure.

  b Name the functional group circled on the structure above. [1]

  c PET belongs to a group of polymers known as condensation polymers.

  Explain the term *condensation polymer*. [3]

> ► Explain the term 'polymer' as well as 'condensation'.

  d PET is made from two monomers, ethane-1,2-diol and benzene-1,4-dicarboxylic acid. Complete the table below by drawing the full structural formula of each of the monomers.

> ► Draw every atom and every bond!

Full structural formula of ethane-1,2,-diol	Full structural formula of benzene-1,4-dicarboxylic acid

[3]

Plastics made from PET are very strong. They are used to manufacture thin-walled tubing for use in medical products.

  e i What is the **strongest** type of intermolecular force which exists between the polymer chains of PET? [1]

> ► Work out all the types of intermolecular forces present before deciding.

   ii Complete the diagram below to show clearly how this type of intermolecular force results between the polymer chains:

   • draw dotted lines to show where these intermolecular forces occur;

   • label the structures to show how these intermolecular forces arise. [3]

> ► You need to show polar bonds using δ+ and δ–.

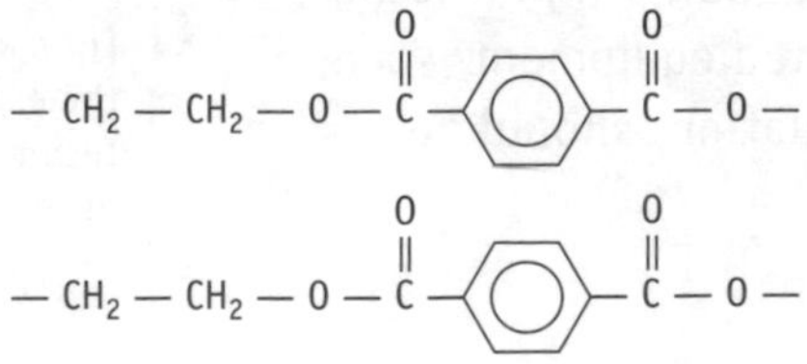

A polymer related to PET, **polymer X**, has been produced. A repeating unit of this is shown below.

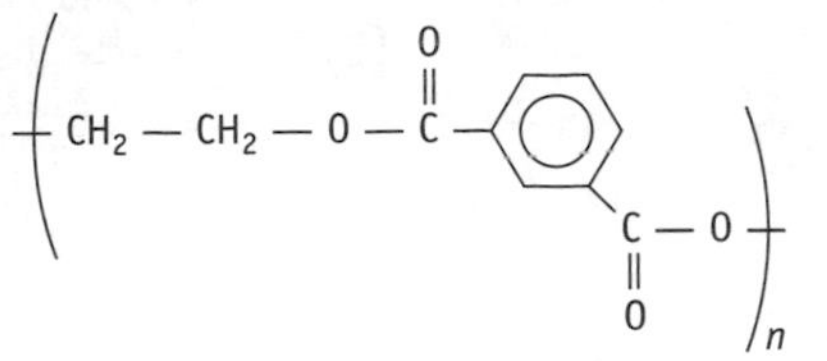

Polymer X

f Explain why the melting temperature of PET (around 600°C) is higher than the melting temperature of **polymer X** (around 209°C). [4]

From OCR January 2002, Module 2853 (now included in 2849) [Total: 16]

> Relate the melting point to the strength of intermolecular forces and the ability of chains to pack together.

2 Aspartame is sold under the trade name *Nutrasweet*. Its sweetness was discovered accidentally by a chemist trying to make an anti ulcer drug. He forgot to wash his hands and noticed a sweet taste on his fingers.

a The chemist had a sample of impure solid aspartame. Describe how he could produce a pure sample of aspartame by recrystallisation using methanol as a solvent. *In this question one mark is available for quality of written communication.* [5]

> Plan your answer before starting to write. Write in complete sentences and include appropriate chemical terms.

The structure of aspartame is shown below.

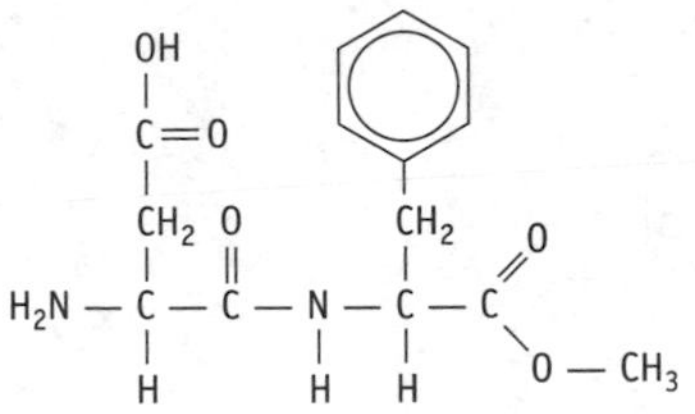

b Clearly circle the following functional groups on the structure of aspartame.

The **amide** group. Label this X.

The **carboxylic acid** group. Label this Y. [2]

> Make sure you label the functional groups clearly.

Aspartame is a chiral molecule. It contains two chiral carbon atoms.

c i On the structure of aspartame above identify each of the two chiral carbon atoms with an asterisk*. [2]

ii What feature makes a carbon atom chiral? [1]

Chemists believe that the sensation of sweetness occurs when the aspartame molecule forms a hydrogen bond with protein molecules in the surface of the tongue.

d On the diagram below circle all the hydrogen atoms on the aspartame molecule which could form hydrogen bonds. [3]

> Three marks available.... Can you find three hydrogen atoms capable of hydrogen bonding?

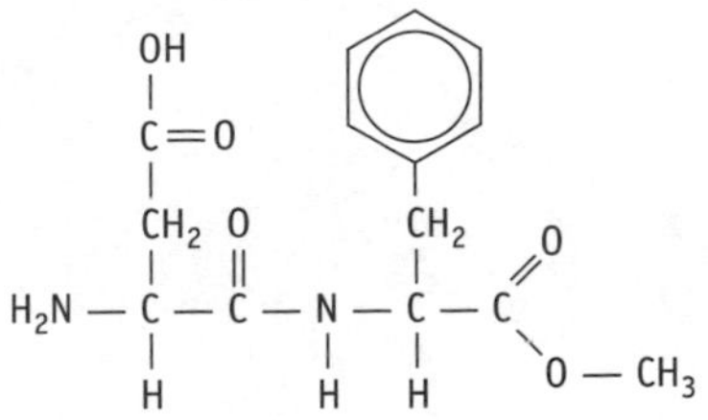

Aspartame exists completely as zwitterions.

e Complete the structure below to show the zwitterion formed by aspartame.

[2]

> ▶ Show any charges clearly.

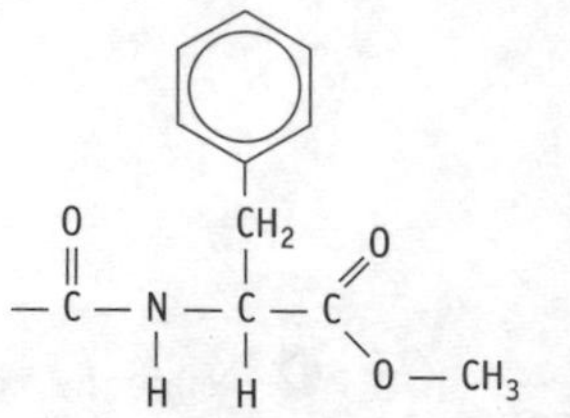

A sample of aspartame can be hydrolysed to give the amino acids shown below.

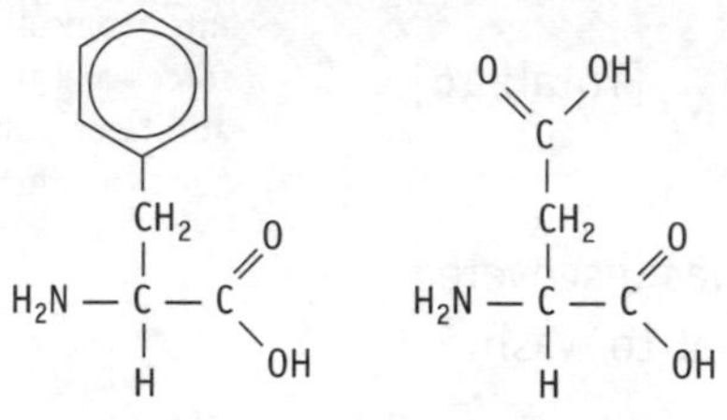

phenylalanine   aspartic acid

f Describe how you could hydrolyse a sample of aspartame. [2]

> ▶ Include the names of any reagents used and essential reaction conditions.

A different dipeptide linkage is formed when the NH_2 group on the aspartic acid reacts with the –COOH group on phenylalanine.

g Draw the structure of the resulting molecule. [2]

From OCR January 2002, Module 2853 (now included in 2849)   [Total: 19]

3 Local anaesthetics such as benzocaine are used to provide a temporary loss of pain to areas of the body. They work by interrupting signals sent along the nerves to the brain.

Benzocaine can be prepared in the laboratory from 4-nitrobenzoic acid in two steps. These are shown below.

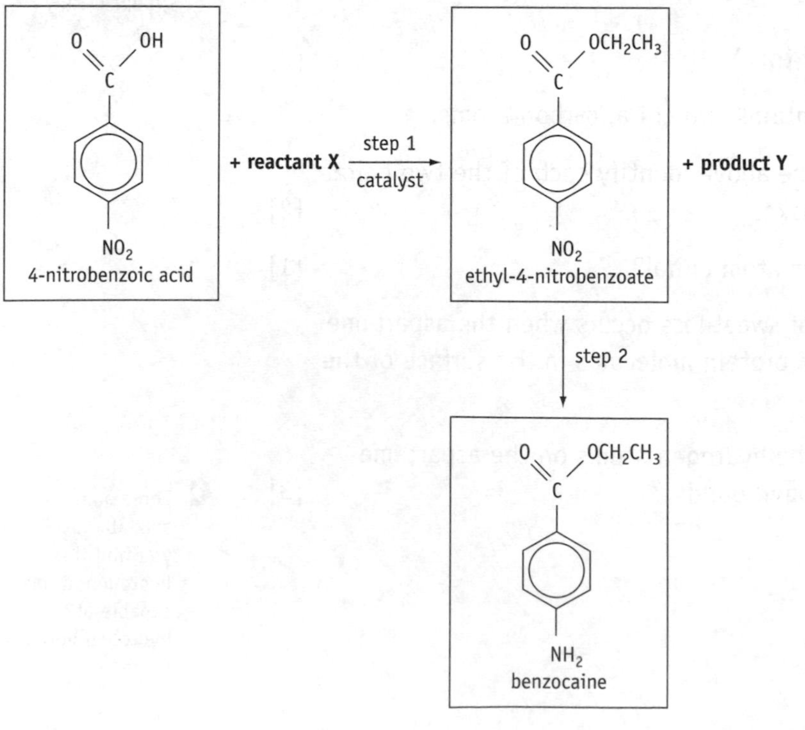

a **i** Draw the structures of **reactant X** and **product Y** in step 1 in the boxes below.

Use the question as a guide to how much detail is required.

reactant X	product Y

[2]

ii Name a suitable catalyst for the reaction in step 1. [1]

Write a name – not a formula!

Compound Z can be used in place of 4-nitrobenzoic acid in step 1. **Compound Z** contains an acyl chloride group and reacts much faster than 4-nitrobenzoic acid to give ethyl-4-nitrobenzoate.

b Draw the structure of **compound Z** showing the full structural formula of the acyl chloride group. [2]

When benzocaine is hydrolysed an alcohol with M_r 46 is isolated. The identity of this alcohol is confirmed using proton n.m.r.

The n.m.r. spectrum is shown below.

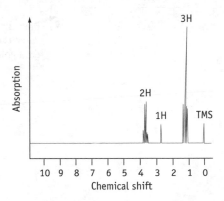

c **i** Use the spectrum above <u>and the data sheet</u> which accompanies this paper to determine the type of proton responsible for each of the chemical shifts.

Make sure you copy very carefully from the data sheet.

Refer to page 22 of this guide for data.

chemical shift from spectrum	relative number of protons	type of proton
1.2	3	
2.7	1	
3.8	2	

[3]

ii Use the information to suggest the **full structural formula** of this alcohol. [1]

Draw every atom and every bond!

From OCR January 2002, Module 2583 (now included in 2849) [Total: 9]

4 Food chemists divide sweeteners into two categories: bulk sweeteners and intense sweeteners. Aspartame is an intense sweetener and sucrose is a bulk sweetener.

Sucrose can be hydrolysed in the presence of hydrogen ions to form glucose and fructose:

$$C_{12}H_{22}O_{11} \ + \ H_2O \ \rightarrow \ C_6H_{12}O_6 \ + \ C_6H_{12}O_6$$
$$\text{sucrose} \qquad\qquad \text{glucose} \qquad \text{fructose}$$

A student carried out this reaction at 25°C in the presence of 1.0 mol dm^{-3} aqueous hydrochloric acid.

The following data were obtained:

time /seconds	concentration of sucrose/mol dm^{-3}
0	1.04
900	0.74
1800	0.52
2700	0.40
3600	0.30
4500	0.22
5400	0.16

a i Plot the data on the axes below.

> Draw a smooth curve through the data points.

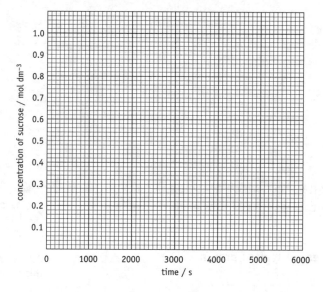

[3]

ii Label two half-lives on your graph. [2]

> Make your labelling very clear.

iii Give the value of each half life.

first half-life .... second half-life .... [2]

> Don't forget the units!

iv What is the order of reaction with respect to sucrose? Explain clearly how you arrived at your answer. [2]

v Explain how you would use your graph to find the initial rate (rate at $t = 0$ s) of the reaction. [2]

In a separate set of experiments, the student found that the reaction is first order with respect to the concentration of hydrogen ions.

b Use the above information and your answer to (a)(iv) to construct the rate equation for the hydrolysis of sucrose in the presence of hydrogen ions. [3]

> Don't mix up your K's!

c What would happen to the rate of reaction if the concentration of the hydrochloric acid was halved? Assume that the concentration of sucrose remains constant. [1]

From OCR January 2002, Module 2853 (now included in 2849) [Total: 15]

5 A hoard of bronze axes was found in Yearsley Common in Yorkshire. They are believed to date back to the Bronze Age (around 2000 BC) when bronze was used to make tools.

Bronze is an alloy of copper and tin and is much stronger than copper alone.

a Explain in terms of the arrangement of atoms how alloying increases the strength of copper. [3]

> Think about how to obtain three separate marks for this answer – a diagram would help.

A chemist determined the amount of copper in a sample of bronze by the following method.

- 0.200 g of bronze were treated with concentrated nitric acid.

- A precipitate of tin(IV) oxide formed and this was filtered off.

- The remaining blue solution contained $Cu^{2+}(aq)$ ions and was made up to 100 cm³ with water in a volumetric flask **(solution P)**.

- The concentration of $Cu^{2+}(aq)$ ions in **solution P** was determined using a colorimeter.

The chemist produced the calibration graph shown below.

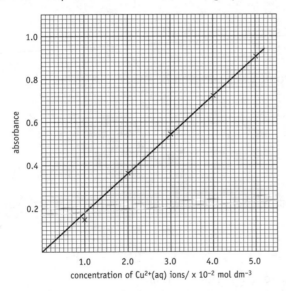

b Outline the steps the chemist would take to produce a calibration graph to use in the determination. [4]

> Explain how many different solutions you would make up and the range of concentrations.

Solution P was found to given an absorbance reading of 0.48.

c i What is the concentration of the $Cu^{2+}(aq)$ ions in mol dm⁻³ in **Solution P**? [1]

> Read the value carefully from the graph.

ii Calculate the mass of copper present in the sample of bronze (A_r: Cu, 63.5) [3]

> Set your calculation out carefully. Give your answer in g to an appropriate number of significant figures.

iii Use your answer from (ii) to calculate the percentage of copper present in the sample of bronze. [1]

From OCR January 2002, Module 2853 (now included in 2849) [Total: 12]

> Show your working!

6 In 1826 John Fredrick Daniell constructed a simple cell.

He used a zinc rod immersed in zinc(II) sulphate solution and a copper can containing copper(II) sulphate solution. The two electrodes were separated by a porous partition.

A diagram of the cell is shown below.

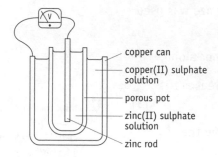

copper can

copper(II) sulphate solution

porous pot

zinc(II) sulphate solution

zinc rod

a Use the information below to (i) write a balanced equation for the overall cell reaction and (ii) calculate $E_{cell}^{\ominus}$.

$Zn^{2+} + 2e^- \rightarrow Zn \quad E^{\ominus} = -0.76 \text{ V}$

$Cu^{2+} + 2e^- \rightarrow Cu \quad E^{\ominus} = +0.34 \text{ V}$

i Equation [1]

ii $E_{cell}^{\ominus}$ = ... V [1]

The atomic number of an atom of copper is 29. In the solution of copper(II) sulphate the oxidation state of the copper is +2.

b i Complete the space below to show the electronic configuration of an **atom** of copper.

$1s^2 2s^2 2p^6 3s^2 3p^6$ [2]

ii State the number of electrons present in the 3d subshell of a **Cu^{2+}** ion.
[1]

Copper(II) sulphate is one of the electrolytes in the Daniell cell. The solution contains the blue complex ion $[Cu(H_2O)_6]^{2+}$.

c Explain what is meant by the term *complex ion*. [2]

The five d orbitals in an isolated Cu^{2+} ion have the same energy. The diagram shows what happens to the energies of these orbitals in the complex ion $[Cu(H_2O)_6]^{2+}$.

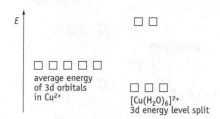

E

average energy of 3d orbitals in Cu^{2+}

$[Cu(H_2O)_6]^{2+}$ 3d energy level split

d Use the diagram above and your understanding of electron energy levels to explain why transition metal compounds such as aqueous copper(II) sulphate are coloured. [4]

> Adding some labels to the above diagram may help you to explain your answer.

e When concentrated hydrochloric acid is added to the copper(II) sulphate solution, the following reaction takes place.

$$[Cu(H_2O)_6]^{2+}(aq) + 4Cl^-(aq) \rightleftharpoons [CuCl_4(H_2O)_2]^{2-}(aq) + 4H_2O(l)$$
$$\text{blue} \qquad\qquad\qquad\qquad \text{yellow}$$

The final solution appears green. What does this tell you about the position of equilibrium? [1]

f Write an expression for the equilibrium constant, K_c, for this reaction. [3]

> Remember $\dfrac{\text{products}}{\text{reactants}}$

The equilibrium constant in (e) is called the stability constant K_{stab} $[CuCl_4(H_2O)_2]^{2-}$. The table below gives information about the stability constants K_{stab} of three complex ions of copper.

complex ion	colour	lg K_{stab}
$[CuCl_4(H_2O)_2]^{2-}$	yellow	5.6
$[Cu(NH_3)(H_2O)_2]^{2+}$	violet	13.1
$[Cu(edta)]^{2+}$	pale blue	18.1

g Use the information in the table to predict what you would expect to **see** when

  i a solution containing edta^{4-} ions is added to the green solution in (e) until just in excess. Explain your answer. [2]

  ii concentrated ammonia solution is added to the solution formed in (g)(i). Explain your answer. [2]

> Say what you would *see* (use one of the colours from the table above) and then follow with an explanation.

From OCR January, 2002, Module 2853 (now included in 2849) [Total: 19]

Module 2854

7 Lactic acid is formed in sour milk. It also occurs in the muscle tissues of athletes suffering from cramp. Its structure is shown below.

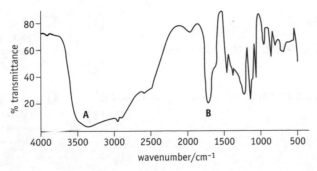

lactic acid

a **i** Give the systematic name for lactic acid. [2]

> Remember to include the carbon in the carboxyl group.

  ii The infra-red spectrum of lactic acid is shown below. (Data about characteristic infra-red absorptions are given in the *Data Sheet* which accompanies this paper.)

> Refer to page 7 of this guide for data.

Which bonds are responsible for the absorptions at **A** and **B**? [2]

> Draw out the bonds.

b Lactic acid can be made industrially by the following route, using ethene as a feedstock.

stage 1 $CH_2 = CH_2 + \frac{1}{2}O_2$ $\xrightarrow{PdCl_2/CuCl_2(aq)}$ $CH_3 - C\overset{\displaystyle O}{\underset{\displaystyle H}{<}}$

ethanal

stage 2 $CH_3 - C\overset{\displaystyle O}{\underset{\displaystyle H}{<}}$ + HCN $\longrightarrow$ **compound A**

stage 3   **compound A** $\xrightarrow{reflux/dilute\ HCl(aq)}$ $CH_3 - \overset{\displaystyle OH}{\underset{\displaystyle H}{\overset{\displaystyle |}{\underset{\displaystyle |}{C}}}} - C\overset{\displaystyle O}{\underset{\displaystyle OH}{<}}$

lactic acid

> **i** What is meant by the term *feedstock*? [1]
>
> **ii** Name a raw material from which ethene can be obtained. [1]

This section is covered in the Visiting the Chemical Industry unit.

c **i** Draw a dot–cross diagram for an ethanal molecule. (Show outer shell electrons only.)

Don't forget to show lone pairs of electrons.

ii Explain why the $- C\overset{\displaystyle O}{\underset{\displaystyle H}{\diagdown}}$ bond angle in ethanal is 120°. [3]

iii Sketch the low resolution n.m.r. (^{1}H) spectrum of ethanal, showing the relative intensities. Use the *Data Sheet* provided. [3]

Remember – chemical shift indicates the chemical environment and the relative heights of the peaks indicate how many protons are in that environment.

Refer to page 22 of this guide for data.

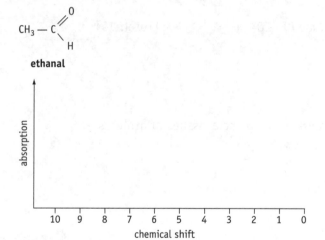

$CH_3 - C\overset{\displaystyle O}{\underset{\displaystyle H}{<}}$

ethanal

d In **stage 2** of the industrial manufacture of lactic acid, ethan**al** reacts with hydrogen cyanide to form a nitrile compound, **compound A**.

$CH_3 - C\overset{\displaystyle O}{\underset{\displaystyle H}{<}}$ + HCN $\longrightarrow$ **compound A**

> **i** Draw the structure of **compound A**. [1]

Make sure you show clearly where bonds have formed.

> **ii** Name the **type** of mechanism for the reaction that produces **compound A**. [2]

> **iii** Draw a diagram to show the first step of this mechanism, the attack of CN⁻ ions on ethan**al**. [3]

Use an arrow to represent the movement of two electrons.

$CH_3 - \overset{\displaystyle OH}{\underset{\displaystyle H}{\overset{\displaystyle |}{\underset{\displaystyle |}{C}}}} - C\overset{\displaystyle O}{\underset{\displaystyle OH}{<}}$

lactic acid

e i Lactic acid exists as two optical isomers. Draw 3-dimensional structures to show these two isomers. Draw a ring round the chiral centre on one of your structures. [3]

Use 3D notation and draw as mirror images.

mirror

ii Lactic acid isolated from muscle tissue exists as a **single** optical isomer. Suggest why. [2]

f A sample of lactic acid is synthesised in the laboratory from ethan**al**. A student started with 5.00 g of ethan**al** (M_r 44.0) and obtained 7.02 g of lactic acid (M_r 90.0). Calculate the % yield of lactic acid. [3]

% yield = $\dfrac{\text{actual yield}}{\text{theoretical yield}} \times 100$

$$CH_3 - C{\overset{O}{\underset{H}{\diagup\!\!\backslash}}} \longrightarrow CH_3 - \overset{\overset{OH}{|}}{\underset{\underset{H}{|}}{C}} - C{\overset{O}{\diagdown OH}}$$

1 mole   1 mole

From OCR June 2002, Module 2854 [Total: 29]

8 Eye-wash solutions often contain boric acid and its salts. These are used to buffer the solution.

a Boric acid is $B(OH)_3$ $\overset{O-H}{\underset{H}{\diagup}} O - B \overset{O-H}{\underset{O-H}{\diagdown}}$

You will have met oxidation states in the AS units. Always give the plus or minus sign.

  i Give the oxidation state of boron in boric acid [1]

  ii Explain why the presence of O–H bonds in boric acid might lead to acidity. [Electronegativity values: O, 3.4; B, 2.0; H, 2.2] [2]

Work out bond polarities.

b Boric acid reacts with water to form an acidic solution. Some books give the following equation.

$$H_3BO_3(aq) \rightleftharpoons H_2BO_3^-(aq) + H^+(aq) \qquad \textbf{equation 2.1}$$

  i Boric acid is described as a weak acid. Explain, by reference to the above reaction, what is meant by a *weak* acid. [1]

Give the definition of a weak acid.

  ii Use **equation 2.1** to write an expression for the acidity constant, K_a, for boric acid. [2]

Remember $\dfrac{\text{products}}{\text{reactants}}$
Remember $pH = -\log[H^+]$

  iii The acidity constant for boric acid from **equation 2.1** is 5.8×10^{-10} mol dm^{-3}. Calculate the pH of 0.10 mol dm^{-3} aqueous boric acid. [3]

You only need to include state symbols where they are asked for, like in this question.

  iv Write an equation for the reaction of aqueous boric acid with aqueous sodium hydroxide, **showing state symbols**. [3]

c i The manufacturers of an eye-wash solution wish to produce a buffer solution to use in their product. They mix a solid salt containing $H_2BO_3^-$ ions with 1.0 dm^3 of 0.10 mol dm^{-3} aqueous boric acid. Calculate the amount in moles of $H_2BO_3^-$ ions they would need to use to achieve a pH of 8.5.

Refer back to page 65 if you can't remember how to do this.

For a buffer solution, $K_a = [H^+] \times [\text{salt}]/[\text{acid}]$ [4]

Suggest means give a sensible reason.

  ii Suggest why a buffer solution is necessary in an eye-wash solution. [3]

iii In this question, one mark is available for the quality of written communication. Explain how a buffer solution based on boric acid works. [5]

From OCR January 2003, Module 2854 [Total: 24]

> To gain the QWC mark, you must write in full sentences, with no bullet points. You must use scientific terminology correctly, and use correct spelling, punctuation and grammar.

9 The **cumene process**, shown below, produces two important industrial chemicals, phenol and propanone. Both compounds are used in the manufacture of plastics.

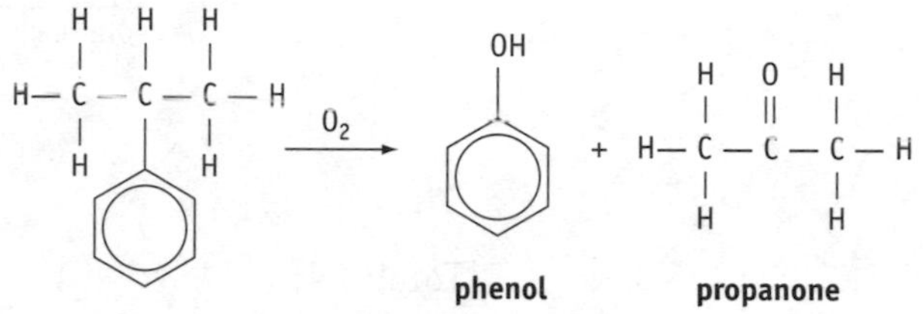

Propanone is also an important solvent.

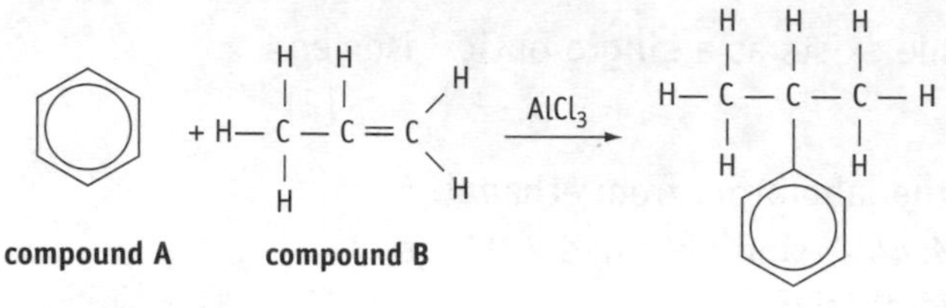

a Give the **names** of compound **A** and compound **B** in **Stage 1** of the **cumene process**. [2]

> Give names – not formulae.

b Suggest **one** advantage and **one** disadvantage of having two useful co-products in an industrial process. [2]

> This is a good example of a Visiting the Chemical Industry question.

c **Stage 1** is a type of Friedel-Crafts alkylation reaction.

  i Choose and underline **two** words from the list below to describe the reaction mechanism involved in a Friedel-Crafts alkylation reaction. [2]

  nucleophilic   radical   addition   substitution   oxidation   electrophilic

> It will be nucleophilic, electrophilic or radical, followed by addition, substitution or oxidation.

  ii From your knowledge of the role of aluminium chloride, $AlCl_3$, in a Friedel-Crafts alkylation reaction, suggest a description of the mechanism involved in **Stage 1** of the **cumene process**. [3]

> You need to write an extended answer for this – think of three possible mark points.

d Propanone can also be produced from propan-2-ol. The alcohol vapour is passed over a heated copper catalyst. The propanone vapour is cooled and condensed to separate it from the hydrogen gas.

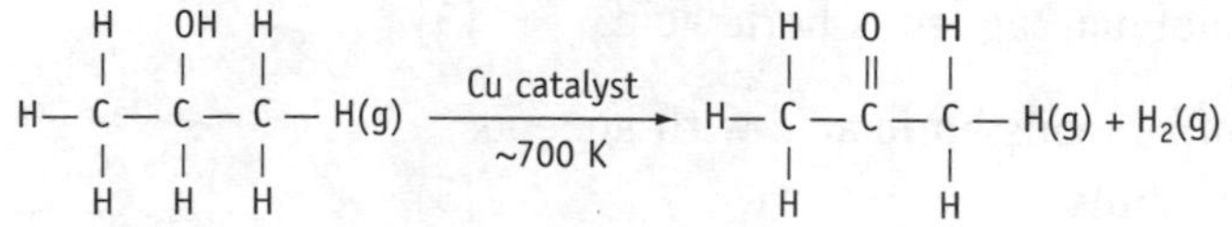

  i Draw a labelled diagram of the laboratory apparatus which could be used to safely carry out a small scale preparation of propanone from propan-2-ol using this method. [4]

> When drawing laboratory apparatus, think of what needs to be collected. Consider safety and use realistic laboratory equipment.

  ii The oxidation of propan-2-ol to propanone in the laboratory is usually carried out by a different method. State the reagents and conditions used. [3]

iii   What colour <u>change</u> would you expect to **see** if you carried out the oxidation of propan-2-ol by the method you have given in **(b)(ii)**?   [2]

Give colours before and after the reaction.

The infra-red spectrum, low resolution n.m.r. (^{1}H) spectrum and the mass spectrum of propanone are shown below.

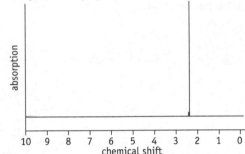

Infra-red spectrum of propanone

propanone

Refer to page 7 of this guide for data on ir.

Refer to page 22 of this guide for data on nmr.

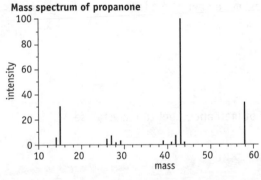

N.m.r spectrum of propanone

Mass spectrum of propanone

Refer to data on the sheet provided in the exam. One piece of evidence is a reading from the spectrum, followed by an interpretation of that reading. Be sure to do this for each of the three spectra.

e   Give and explain **one** piece of evidence from **each** spectrum which would help to identify propanone. Use the *Data Sheet* provided.   [6]

f   The solvent MIBK (methylisobutylketone) can be made from propanone.

MIBK

i   It is difficult to use the i.r. spectrum of MIBK to distinguish it from the starting material propanone. Explain why.   [2]

Think about functional groups.

ii   Name another analytical technique which could be used to distinguish between propanone and MIBK.   [1]

Think through the techniques you have met so far on this course.

From OCR January 2002, Module 2854   [Total: 27]

Mark schemes – Module 2849

Question	Expected answers	Marks
1 a	To prevent oxidation of food/food decomposition/food spoilage/food contamination; to prevent escape of CO_2/gas escaping	1
1 b	Ester	1
1 c	Many monomers joined together; small molecule is eliminated	3
1 d	Ethane-1,2-diol Benzene-1,4-dicarboxylic acid 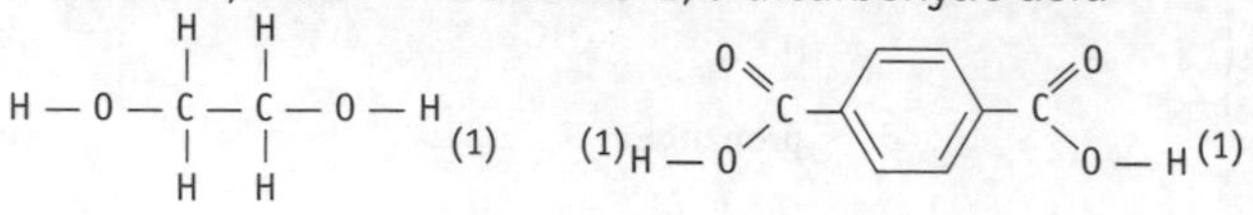	3
1 e i	Permanent dipole – permanent dipole forces	1
1 e ii	δ^- on O on C=O on one chain; δ^+ on C of carbonyl group; attraction shown clearly between these atoms on separate chains; 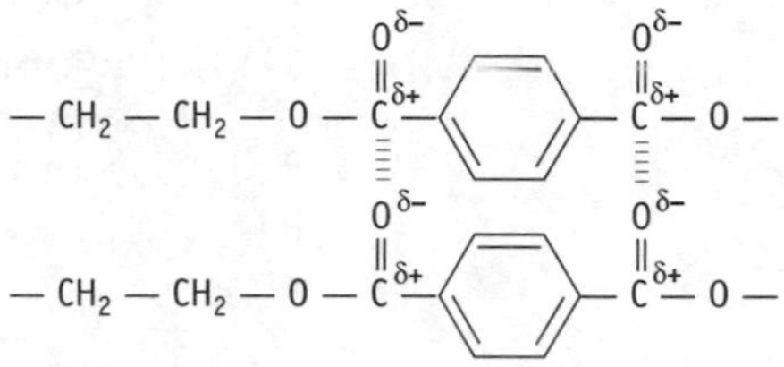 Ignore extra correct charges	3
1 f	Four from: chains are more linear in PET; therefore able to pack more closely together; PET is more crystalline; stronger intermolecular forces; more energy needed to separate the chains	4
	Total	16

Question	Expected answers	Marks
2 a	Three from: dissolve in minimum amount; of hot methanol; cool to recrystallise. One from: filter/dry/wash	4(+1)
2 b	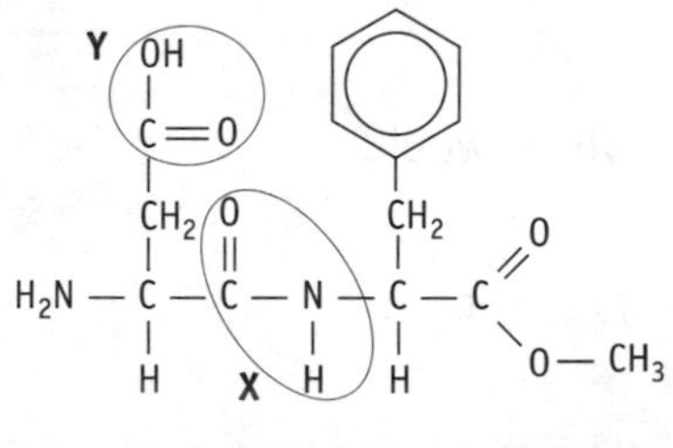 Correctly labelled (i.e. no labels scores 1)	2
2 c i	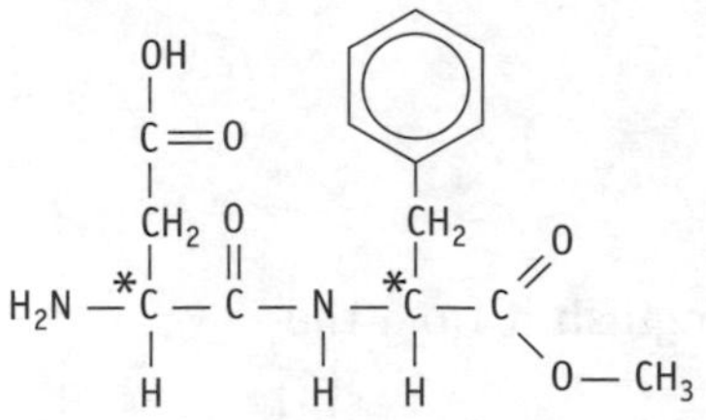	2
2 c ii	Four different groups attached to the carbon atom	1

Question	Expected answers	Marks
2 d	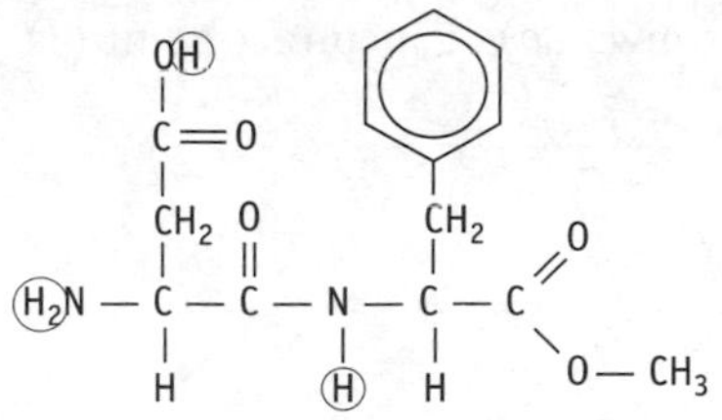 (structure: dipeptide methyl ester with OH and H and H₂N circled) Deduct 1 mark for each additional H circled above 3 (max. 3)	3
2 e	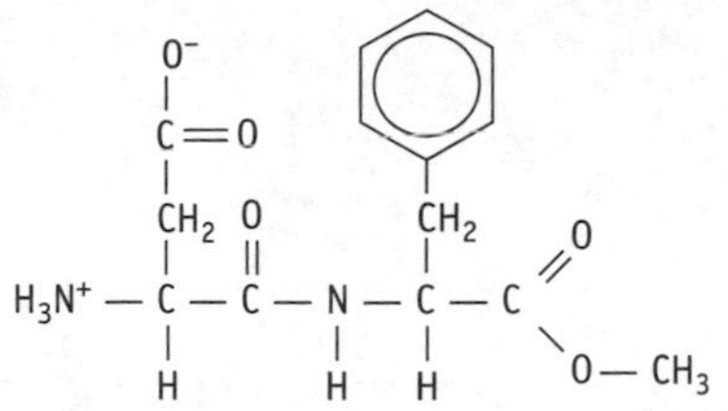 (structure: zwitterion form with O^-, H_3N^+) Correct zwitterion without charges scores 1	2
2 f	Reflux; with moderately concentrated acid/alkali	2
2 g	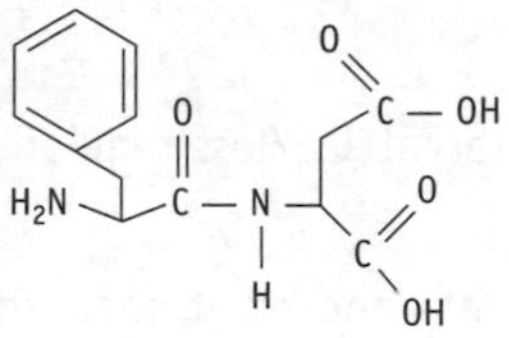 (structure: amine-containing dicarboxylic acid dipeptide fragment)	2
	Total	19

Question	Expected answers	Marks
3 a i	Compound X is $\quad\quad C_2H_5OH$ (1) Other product is $\quad\quad H_2O$ (1)	2
3 a ii	concentrated sulphuric acid	1
3 b	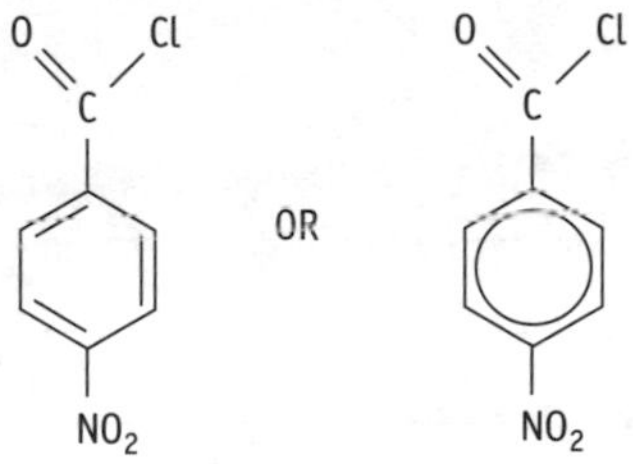 (structure: acid chloride benzene ring with NO_2) OR (structure: acid chloride benzene ring with NO_2)	2
3 c i	R-C$\underline{H}_3$ = 1.2 R-O$\underline{H}$ = 2.7 R-C$\underline{H}_2$-O = 3.8	3
3 c ii	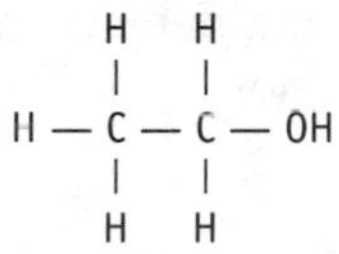 (structure: H–C–C–OH ethanol skeletal with H atoms)	1
	Total	9

Question	Expected answers	Marks
4 a i	Correctly plotted data (2) (all points should be +/– 1 scale division), 1 incorrect point (1) smooth curve	3
4 a ii	Each half-life correctly drawn and **clearly labelled** 1 mark each	2
4 a iii	1800–2200 s	2
4 a iv	First order wrt sucrose; half-lives (almost) constant.	2
4 a v	Draw a tangent to the curve at $t = 0$ s; find the gradient, i.e. concentration ÷ time	2
4 b	Rate = k[H$^+$] [sucrose]; 1 mark for each correct component	3
4 c	Rate would halve	1
	Total	15

Question	Expected answers	Marks
5 a	Different sized atoms; interrupt orderly arrangement of atoms in metal lattice/less regular lattice; layers of atoms prevented from slipping	3
5 b	Four from: make up solutions of known concentrations; range/different concentrations; range in correct region; choose filter/set colorimeter at the correct wavelength; zero with water; measure absorbance/transmittance	4
5 c i	2.65×10^{-2} to 2.70×10^{-2} (mol dm^{-3})	1
5 c ii	$2.65 \times 10^{-2} \times 63.5(1) = 1.68$ g in 1 dm^{-3}; $2.70 \times 10^{-2} \times 63.5 = 1.71$ g; answer ÷ 10 (1) 0.168 g – 0.171 g in sample (1) 2/3 sig fig	3
5 c iii	$\dfrac{0.168}{0.200} \times 100 = 84.0\%$ to $\dfrac{0.171}{0.200} \times 100 = 85.5\%$	1
	Total	12

Question	Expected answers	Marks
6 a i	$Cu^{2+} + Zn \rightarrow Zn^{2+} + Cu$	1
6 a ii	+1.1 V	1
6 b i	$3d^{10}$; $4s^1$	2
6 b ii	9	1
6 c	(central) metal ion; surrounded by/bonded to ligands	2
6 d	Four from: ligands cause splitting of subshell into two energy levels; d orbitals partially filled; electron is promoted/excited from lower energy level to a higher energy level; difference in energy corresponds to visible region of EMS/light (energy) is absorbed; light not absorbed/transmitted/reflected gives colour	4
6 e	Equilibrium lies neither to the left or right	1
6 f	$Kc = \dfrac{[CuCl_4(H_2O)_2{}^{2-}]([H_2O]^4)}{[Cu(H_2O)_6{}^{2+}][Cl^-]^4}$ top component = 1 (with or without ([H$_2$O])^4) bottom component = 1 K_c and powers = 1	3

Question	Expected answers	Marks
6 g i	solution would turn pale blue; stability constant is greater for edta^{4-} complex/edta^{4-} complex is more stable than $[CuCl_4(H_2O)_2]^{2+}$	2
6 g ii	solution would remain pale blue; stability constant for ammonia complex is smaller	2
	Total	19

Mark schemes – Module 2854

Question	Expected answers	Marks
7 a i	2-hydroxypropanoic acid	2
7 a ii	(A) O-H (allow OH but not name) (B) C=O (don't allow CO or name)	2
7 b i	Reactant (in a chemical process)/starting material	1
7 b ii	Crude oil/(cracking) polymers/coal/alkanes/ethane/hydrocarbons/ethanol	1
7 c i	both lone pairs on oxygen double bond single bonds	3
7 c ii	Three regions of electron density/bonds/groups; electrons repel; as far apart as possible/to minimise repulsion	3
7 c iii	Chemical shifts correct (2.2 and 10); relative intensities correct (3:1); only 2 peaks shown (other than TMS at 0)	3
7 d i	(Note: bonds must be unambiguously in the correct place)	1
7 d ii	Nucleophilic; addition	2
7 d iii	Three out of four for: 1 for each arrow (of any kind) 1 for polarisation 1 for correct intermediate	3
7 e i	Chiral centre correct; tetrahedral appearance, using wedges and dashes; mirror image or other enantiomer 	3
7 e ii	Two out of: molecular fit; receptor; enzymes/isomers are stereospecific/made from chiral precursor	2
7 f	5 g ethanal $= \dfrac{5}{44}$ mol, therefore should produce $\dfrac{5}{44}$ mol lactic acid; $\rightarrow \dfrac{5 \times 90}{44}$ g lactic acid $(= 10.277$ g$)$ % yield $= \dfrac{7.02 \times 100}{10.227} = 68.4$ to 68.6%	4
	Total	30

Question	Expected answers	Marks
8 a i	+3	1
8 a ii	$^{\delta-}O - H^{\delta+}$ polar (or partial charges shown) (1); H^+ formed (1)	2
8 b i	In equilibrium/partial dissociation/ionisation	1
8 b ii	$K_a = [H^+] [H_2BO_3^-]/[H_3BO_3]$ top (1); bottom (1) *missing [ ] scores max 1*	2
8 b iii	$[H^+] = \sqrt{K_a[H_3BO_3]}$ (1); $= 7.6 \times 10^{-6}$ (1); pH = 5.1 (1) *accept "5" if working shown*	3
8 b iv	$H_3BO_3(aq) + NaOH(aq) \rightarrow NaH_2BO_3$ (or ions) (aq) $+ H_2O(l)$ reactants and products (1); balancing (1); state symbols (*provided water formed*) (1) *Accept equations forming other salts.*	3
8 c i	pH 8.5 gives $[H^+] = 3.16 \times 10^{-9}$ (1) *then either:* $\dfrac{[salt]}{[acid]} = \dfrac{5.8 \times 10^{-10}}{3.16 \times 10^{-9}}$ (1) = 0.184 (1) Thus 0.018 mol of $H_2BO_3^-$ must be added (1) *or* $[salt] = K_a \times [acid]/[H^+]$ (1) $= 5.8 \times 10^{-10} \times 0.1/3.16 \times 10^{-9}$ (1) *subsumes last mark* = 0.018 (1)	4
8 c ii	Acid/alkali in eye causes damage/irritation/harm (1) buffers maintain pH/neutralise (1) in presence of (small amounts of) acid/alkali/at/near 8.5/neutral pH/same pH as eye/natural pH (1)	3
8 c iii	Indication that acid is H^+/alkali is OH^- (1) (on adding acid) equilibrium moves to left/buffer accepts H^+/*or equation* (1); (on adding alkali) equilibrium moves to right/forms H^+ to neutralise (1); Because $[H_3BO_3]$ and $[H_2BO_3^-]$ large, pH remains constant (1) *QWC SPAG:* spelling (allow one error), punctuation and grammar correct.	5
	Total	24

Question	Expected answers	Marks
9 a	A = benzene; B = propene	2
9 b	Advantage = sale of co-product adds to profit/less waste to dispose of Disadvantage = separation of co-products is costly/different demands for each product	2
9 c i	Electrophilic; substitution	2
9 c ii	$AlCl_3$ polarises the propene molecule; the $\delta+$ carbon; reacts with the benzene ring (clearly labelled diagram acceptable)	3
9 d i	Tube connected with no leaks; production of vapour; vapour contacts catalyst; liquid product is condensed/$H_2(g)$ collected or vented safely	4
9 d ii	Acidified; potassium dichromate; reflux (allow formulae if correct)	3
9 d iii	Orange; to green-blue	2

Question	Expected answers	Marks
9 e	Infra-red: C=O; at ~ 1700 cm^{-1} n.m.r.: only one signal; implies only one type of ^{1}H; OR peak at 2.2, suggests –CO-CH$_3$ Mass spec: molecular ion peak at 58; matches M_r of propanone (allow loss of –CH$_3$; to give fragment at 43)	6
9 f i	Same type of bonds present so same absorptions; strong absorption at 1700 cm^{-1} for C=O; both have C-H and C=O bonds (2 marks for 2 points made)	2
9 f ii	Mass spec/^{1}H nmr/^{13}C nmr/glc	1
	Total	27

Answers to quick check questions

Module 2849

What's in a Medicine? (WM)

Carboxylic acids

Page 3

1 Hexanoic acid

2 $CH_3CH_2COOH + KOH \rightarrow CH_3CH_2COO^-K^+ + H_2O$

3 Butanoic acid $CH_3CH_2CH_2COOH$ and propan-1-ol $CH_3CH_2CH_2OH$.

The OH group in alcohols, phenols and carboxylic acids, and esters

Page 5

1 (a) A phenol and ester, B phenol and carboxylic acid, C hydroxyl.

   (b) Order of acidity, with weakest acid first, is C, A, B.

   (c) React compound B with anhydrous ethanoic anhydride at room temperature OR react compound B with anhydrous ethanoyl chloride at room temperature.

   (d) Shake the aspirin sample with neutral iron(III) chloride solution. Any purple coloration indicates presence of a phenol group, suggesting not all of the 2-hydroxybenzoic acid has reacted.

Infrared spectroscopy

Page 8

1 (a)

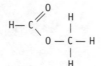

ethanoic acid   methyl methanoate

(b) **Spectrum A**
   • broad absorption at 3200–2800 cm^{-1} suggests O—H bond in carboxylic acid
   • sharp absorption at 1760 cm^{-1} suggests C=O bond

   Spectrum A is ethanoic acid

Spectrum B
   • absence of a broad absorption at 3200–2800 cm^{-1} suggests no O—H group
   • sharp absorption at 1760 cm^{-1} suggests C=O bond

   Spectrum B is methyl methanoate

2 The IR spectrum for salicylic acid has an absorption at 3240 cm^{-1}, due to the presence of the –OH group in a phenol. This is not present in the spectrum of pure aspirin.

Mass spectrometry

Page 11

1

Peak at m/e	Ion responsible
60	CH_3COOH^+
45	$COOH^+$
43	CH_3CO^+
15	CH_3^+

Molecular ion peak is at $m/e = 60$, which suggests the compound is ethanoic acid, CH_3COOH.

2 Compound A must be a secondary alcohol because it oxidises to give only one product, which must be a ketone.

Spectrum A

Peak at m/e	Ion responsible
60	$(CH_3)_2CHOH^+$
59	$(CH_3)_2CHO^+$
45	CH_3CHOH^+

Spectrum B

Peak at m/e	Ion responsible
59	Isotope peak
58	$(CH_3)_2C=O^+$
43	$CH_3C=O^+$
15	CH_3^+

When propan-2-ol (Spectrum A) is oxidised, it produces propanone (Spectrum B).

3

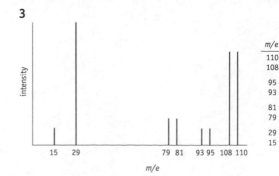

m/e	ion
110	$C_2H_5{}^{81}Br^+$
108	$C_2H_5{}^{79}Br^+$
95	$CH_2{}^{81}Br^+$
93	$CH_2{}^{79}Br^+$
81	$^{81}Br^+$
79	$^{79}Br^+$
29	$CH_3CH_2^+$
15	CH_3^+

Designer Polymers (DP)

The structure and properties of polymers

Page 13

1 (a)

$$-N - (CH_2)_4 - N - C - (CH_2)_5 - C - N - (CH_2)_4 - N -$$
(with H on each N, and O double-bonded to each C)

(b)

$$-O - CH_2CH_2 - O - C - \text{[naphthalene]} - C - O - CH_2CH_2 - O$$
(with O double-bonded to each C)

Amines and amides

Page 15

1 (a) methylamine (b) phenylamine (c) propylamine
   (d) methylpropylamine.

2 (a) products: $CH_3CH_2CH_2 - N - H$ with CH_3 below N + HCl

   (b) products:

   $$H_3C - C - N - CH_2CH_3 + HCl$$
   (with O double-bonded to C, H below N)

3 (a) Conditions: reflux with moderately conc. HCl

   $$CH_3 - C - N - C_2H_5 + H_2O \xrightarrow{H^+} CH_3 - C\begin{smallmatrix}O\\OH\end{smallmatrix} + C_2H_5NH_3^+$$
   (with O double-bonded to first C, H below N)

   (b) Conditions: reflux with moderately conc. NaOH

   $$\left[C - (CH_2)_4 - C - N - (CH_2)_6 - N \right] + 2\,NaOH \longrightarrow$$
   (with O double-bonded to each C, H below each N)

   $$^+Na^-O \begin{smallmatrix}O\\\end{smallmatrix} C - (CH_2)_4 - C \begin{smallmatrix}O\\O^-Na^+\end{smallmatrix} + H_2N - (CH_2)_6 - NH_2$$

Engineering Proteins (EP)

Optical isomerism

Page 17

1

$$H - \overset{CH_3}{\underset{OH}{C}} - CN \quad\quad NC - \overset{CH_3}{\underset{HO}{C}} - H$$

2 (a)

$$\overset{H_3C}{\underset{H_2C}{}}C = \overset{*}{\bigcirc} - CH_3$$

   (b) $CH_3\overset{*}{C}H(NH_2)\,COOH$

Amino acids

Page 19

1

alkaline solution	neutral solution	acidic solution

$$H_2N - \overset{H}{\underset{\underset{H_3C\ CH_3}{CH}}{C}} - C\overset{O}{\underset{O^-}{}}$$
$$H_3\overset{+}{N} - \overset{H}{\underset{\underset{H_3C\ CH_3}{CH}}{C}} - C\overset{O}{\underset{O^-}{}}$$
$$H_3\overset{+}{N} - \overset{H}{\underset{\underset{H_3C\ CH_3}{CH}}{C}} - C\overset{O}{\underset{OH}{}}$$

2

$$H_2N - \overset{H}{\underset{H}{C}} - \overset{O}{C} - N - \overset{CH_2OH}{\underset{H}{C}} - C\overset{O}{\underset{OH}{}}$$
(with H on middle N)

and

$$H_2N - \overset{CH_2OH}{\underset{H}{C}} - \overset{}{C} - N - \overset{H}{\underset{H}{C}} - C\overset{O}{\underset{OH}{}}$$
(with O on middle C, H on middle N)

3 (a)

$$H_2N - \overset{COO^-}{\underset{\underset{H}{CH_2}}{C}} - COO^- \quad and \quad H_2N - \overset{H}{\underset{H}{C}} - COO^- \quad and \quad H_2N - \overset{CH_3}{\underset{H}{C}} - COO^-$$

   (b)

$$H_3N^+ - \overset{COOH}{\underset{\underset{H}{CH_2}}{C}} - COOH \quad and \quad H_3N^+ - \overset{H}{\underset{H}{C}} - COOH \quad and \quad H_3N^+ - \overset{CH_3}{\underset{H}{C}} - COOH$$

Nuclear magnetic resonance (NMR) spectroscopy

Page 23

1 (a) Two peaks with areas in the ratio 3:1

   (b) One peak only

   (c) Two peaks with areas in the ratio 3:1

   (d) Two peaks with areas in the ratio 3:1

2 (a) Three different types of proton

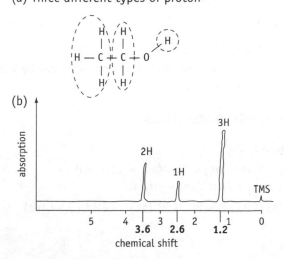

   (b)

3 The molecular formula could be that of an ester or a carboxylic acid. The absorption peak at chemical shift 11.8 suggests the molecule is an acid and the peak at chemical shift 1.2 suggests there are two methyl groups in the molecule. The molecule is 2-methylpropanoic acid.

Equilibria and concentrations

Page 25

1 (a) $K_c = \dfrac{[SO_3]^2}{[SO_2]^2 [O_2]}$ Units = $mol^{-1}\,dm^3$

(b) $K_c = \dfrac{[NO_2]^2}{[N_2O_4]}$ Units = $mol\,dm^{-3}$

2 $K_c = 2.38 \times 10^{-3}\,mol^{-2}dm^6$

3 (a) Increasing the pressure would shift the position of equilibrium to the right. K_c would be unaffected.

(b) Increasing the temperature would shift the equilibrium in the endothermic direction – left. The value of K_c would decrease.

The effect of concentration on rate

Page 28

1 (a) Measure the volume of oxygen evolved at known time intervals.

(b) Measure the amount of bromine produced at given time intervals using a colorimeter.

2 First order wrt BrO_3^-, first order wrt Br^-, second order wrt H^+ and overall order 4.

3 Rate = $k[S_2O_8^{2-}][I^-]$

4 (b) Half-lives should be in the range 170–190 s.

(c) As the half-lives are roughly constant, the reaction is first order wrt ester.

The Steel Story (SS)

Where does colour come from?

Page 30

1 (a) green (b) blue-violet

2 This maximises the amount of light absorbed by the coloured solution.

Electrode potentials

Page 33

1 (a) $2MnO_4^-(aq) + 16H^+(aq) + 10Cl^-(aq) \rightarrow$
$5Cl_2(g) + 2Mn^{2+}(aq) + 8H_2O(l)$

(b) $H_2O_2(aq) + 2H^+ + 2I^-(aq) \rightarrow 2H_2O(l) + I_2(aq)$

2 (a) Make a Cu^{2+}/Cu half-cell by dipping a copper rod into a solution of 1.00 mol dm^{-3} copper sulphate solution. Connect this to a standard hydrogen half-cell (298K, 1 atm pressure) using a salt bridge and a high-resistance voltmeter. Take the reading of the voltmeter, noting which half-cell is connected to the positive terminal.

(b) Make a Cl_2/2Cl^- half-cell by dipping a platinum electrode into a solution containing 1 mol dm^{-3} $Cl_2(aq)$ and 1 mol dm^{-3} $Cl^-(aq)$. Proceed as in (a), connecting to a standard hydrogen half-cell.

3 (a) +1.53 V (b) +0.46 V

4 (a) No reaction occurs.

(b) $Cr_2O_7^{2-}$ oxidises Fe^{2+} to Fe^{3+}.

(c) $Cr_2O_7^{2-}$ oxidises Br^- to Br_2.

The d block: transition metals

Page 35

1 (a) Ti^{2+}: $1s^2 2s^2 2p^6 3s^2 3p^6 3d^2$

(b) Cr^{3+}: $1s^2 2s^2 2p^6 3s^2 3p^6 3d^3$

(c) V^{3+}: $1s^2 2s^2 2p^6 3s^2 3p^6 3d^2$

2 (a) +6 (b) +4 (c) +5

3 Copper is a transition metal because one of its ions (Cu^{2+}) has a partially filled d subshell.

4 (a) Homogeneous

(b) The Co^{2+} ion acts as a catalyst because it is able to change its oxidation state (to Co^{3+}, and then back to Co^{2+}).

Complex formation

Page 37

1 (a) A molecule or anion with a lone pair of electrons, which forms a dative bond with the central metal atom or ion in a complex.

(b) A central metal ion surrounded by ligands.

(c) The number of bonds formed between the central metal ion and the surrounding ligand(s).

(d) A bond where one atom (from the ligand) provides both the electrons.

2 (a) hexaaquachromium(III) ion

(b) tetrachlorocobalate(II) ion

(c) tetraaquadihydroxoiron(III) ion

$$\left[\begin{array}{c} \overset{\displaystyle \ddot{O}H_2}{\underset{\displaystyle \ddot{O}H_2}{\overset{\displaystyle |}{\underset{\displaystyle |}{Fe}}}} \end{array} \right]^{+}$$

$$H_2O\overset{..}{:}\cdots \quad \cdots \overset{..}{:}OH$$
$$H_2O\overset{..}{:} \quad \overset{..}{:}OH$$

3 (a) $K_{stab} = \dfrac{[[Cu(NH_3)_4(H_2O)_2]^{2+}]}{[[Cu(H_2O)_6]^{2+}][NH_3]^4}$

(b) $[Cu(NH_3)_4(H_2O)_2]^{2+}$ has $\log K_{stab} - 12$, making it more stable than $[Cu(H_2O)_6]^{2+}$, for which $\log K_{stab} = 5.6$.

Module 2854

Aspects of Agriculture (AA)

Nitrogen and Group 5

Page 39

1 Plant growth (crops) and leaching.

2 Nitrifying bacteria carry out oxidation as the highest oxidation state (nitrate V) is produced.

Bonding, structure and properties

Page 41

Substance	Structure	State	Solubility (aq)	Electrical conductivity
vanadium	giant metallic	solid	insoluble	high
xenon	simple molecular	gas	insoluble	low
cotton	macromolecular	solid	insoluble	low
potassium iodide	giant ionic	solid	soluble	low except (l) or (aq)
propan-1-ol	simple molecular	liquid	soluble	low
steel	giant metallic	solid	insoluble	high
glass	giant covalent	solid	insoluble	low
polyester	macromolecular	solid	insoluble	low
lead nitrate	giant ionic	solid	soluble	low except (l) or (aq)
silicon carbide	giant covalent	solid	insoluble	low

1 and **2**

Equilibria and partial pressures

Page 42

1 $K_p = \dfrac{p_{CO}p_{H_2}^3}{p_{CH_4}\,p_{H_2O}}$ Units $= \dfrac{atm \times atm^3}{atm \times atm} = atm^2$

2 $K_p = \dfrac{0.20 \times (0.49)^3}{0.20 \times 0.16} = 0.74\ atm^2$

Partition equilibrium

Page 43

1 $K_{ow} = \dfrac{17.3}{0.202} = 85.6$

2 $K_{ow} = \dfrac{0.70}{0.07} = 10$

3 $K_{ow} = \dfrac{[DDT(org)]}{[DDT(aq)]} \quad 9.5 \times 10^5 = \dfrac{[DDT(org)]}{0.002}$

   $[DDT(org)] = 1900 \text{ g dm}^{-3}$

Colour by Design (CD)

Oils and fats

Page 45

1

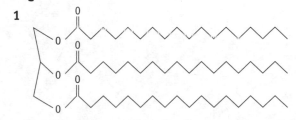

2 Saturated fats are solids because the intermolecular forces are stronger than unsaturated fats due to closer and better packing.

3 Hydrolyse the fat then acidify to liberate the free fatty acids. Run a glc trace.

Ultraviolet and visible spectroscopy

Page 47

1 The pigment is red, because it reflects the wavelengths corresponding to red light and absorbs those corresponding to blue light.

Chromatography

Page 48

1 $R_f = 1.0/2.6 = 0.38$

2 The methyl esters are more volatile than the fatty acids because they will have less affinity for the column. This means traces can be produced more quickly.

Arenes

Page 49

1 (a)   (b)

2 (a) 6 (b) 10

3 According to the diagram 152 kJ mol^{-1}

Reactions of arenes

Page 51

1 Reagents – mixture of conc. sulphuric and conc. nitric acids, warm.

2 Add $AlCl_3$ for chlorine or $FeBr_3$ for bromine.

3

Azo dyes

Page 52

1 (a) $NaNO_2 + HCl \longrightarrow HNO_2 + NaCl$

   (b)

   (c)

2

then

Chemistry of colour

Page 53

1 (a) Ni^{2+}: $1s^2 2s^2 2p^6 3s^2 3p^6 3d^8$

   (b) The complex is green because it absorbs light in the red/violet region of the visible spectrum. When the nickel ion is surrounded by ligands, the 3d subshell is split in energy. As light is absorbed, electrons are excited from the lower to the higher level. The gap in energy between the two levels is related to the frequency of the absorbed light by the equation $\Delta E = h\nu$.

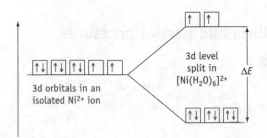

(c) Changing the ligand changes the energy gap (excitation energy) between the two levels.

2 (a) 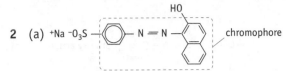 chromophore

(b) The molecule can absorb light because it contains an extended delocalised system of electrons. This molecule absorbs blue light and therefore appears orange.

(c) If functional groups such as $-OH$, $-NH_2$ or $-NR_2$ are attached to the chromophore, this modifies the structure, allowing different wavelengths of light to be absorbed.

The Oceans (O)

Energy, entropy and equilibrium

Page 55

1 $\Delta S_{surr} = -\dfrac{100,300}{773} = -129.8$ J K^{-1}mol^{-1}

$\Delta S_{sys} = +174.8$ J K^{-1} mol^{-1}

$\Delta S_{total} = +174.8 - 129.8 = +45$ J K^{-1} mol^{-1}

Positive total entropy change and so the reaction occurs spontaneously.

2 For the reaction to just occur

$\Delta S_{total} = \Delta S_{sys} + \Delta S_{surr} = 0$

$0 = +174.8 - \dfrac{100,300}{T}$ $T = \dfrac{100,300}{174.8} = 574$ K

Energy changes in solution

Page 57

1 Least to most negative ΔH_{LE}: KF, LiF, CaF$_2$, CaO.

2 Least to most negative ΔH_{hyd}: K$^+$, Na$^+$, Ca^{2+}, Mg^{2+}, Al^{3+}.

3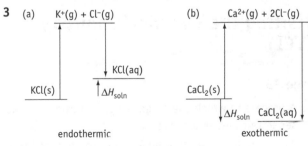

4 $\Delta H_{solution} = -\Delta H_{LE} + \Delta H_{hyd(Ag^+)} + \Delta H_{hyd(I^-)}$
   $= +802 - 446 - 293 = +63$ kJ mol^{-1}

5 AgI is insoluble in water because $\Delta H_{solution}$ is too endothermic.

Born–Haber cycles

Page 59

1 Lattice enthalpy of sodium chloride:

$\Delta H_f(NaCl) = \Delta H_{at}(Na) + \Delta H_{i(1)}(Na) + \Delta H_{at}(Cl) + \Delta H_{EA(1)}(Cl) + \Delta H_{LE}$

$-411 = +107 + 502 + 121 + (-355) + \Delta H_{LE}$

$\Delta H_{LE} = -411 - 107 - 502 - 121 + 355 = -786$ kJ mol^{-1}

Lattice enthalpy of calcium chloride:

$\Delta H_f(CaCl_2) = \Delta H_{at}(Ca) + \Delta H_{i(1)}(Ca) + \Delta H_{i(2)}(Ca) + 2\Delta H_{at}(Cl) + 2\Delta H_{EA(1)}(Cl) + \Delta H_{LE}$

$-796 = +178 + 596 + 1152 + 242 + (-710) + \Delta H_{LE}$

$\Delta H_{LE} = -796 - 178 - 596 - 1152 - 242 + 710$
$\quad = -2254$ kJ mol^{-1}

2 The lattice enthalpy for calcium chloride is much larger, as the attraction between the ions is greater, as a calcium ion is smaller and has a 2+ charge.

3
```
                    K⁺(g) + Cl(g)                                    −355 kJ mol⁻¹
+425 kJ mol⁻¹ │ΔH°ᵢ(₁)(K)              ΔH°ₑₐ(₁)(Cl)
                                                    K⁺(g) + Cl⁻(g)
        K(g) + Cl(g)
+121 kJ mol⁻¹ │ΔH°ₐₜ(Cl)
        K(g) + ½Cl₂(g)                          ΔH°ₗₑ(KCl)
+89 kJ mol⁻¹ │ΔH°ₐₜ(K)                          −711 kJ mol⁻¹
        K(s) + ½Cl₂(g)
        −431 kJ mol⁻¹ │ΔH°f(KCl)                  KCl(s)
```

Solubility products

Page 61

1 A saturated solution is a solution containing the maximum concentration of solute at the given temperature.

2 The common ion effect is when the solution already contains one of the ions in the solute.

3 $K_{sp} = [Pb^{2+}(aq)][Cl^-(aq)]^2$

4 $K_{sp} = [Ag^+(aq)] [I^-(aq)] = 8 \times 10^{-10}$ mol^2 dm^{-6}

$[AgI(aq)] = [Ag^+(aq)] = \sqrt{8 \times 10^{-10}} = 2.83 \times 10^{-5}$ mol dm^{-3}

in g dm^{-3} $[AgI(aq)] = 2.83 \times 10^{-5} \times (107.9 + 126.9)$
$\quad\quad\quad\quad\quad\quad = 6.64 \times 10^{-3}$ g dm^{-3}

$[Ca^{2+}(aq)][CO_3^{2-}(aq)] = 0.0001 \times 0.0005$
$\quad\quad\quad\quad\quad\quad\quad\quad = 5 \times 10^{-8}$ mol^2 dm^{-6}

This exceeds K_{sp} so a precipitate would form.

$Ca^{2+}(aq) + CO_3^{2-}(aq) \rightarrow CaCO_3(s)$

Weak acids and pH

Page 63

1 pH = 3

2 $[OH^-] = 0.05$ mol dm^{-3} $K_w = [H^+][OH^-] = 10^{-14}$ mol^2 dm^{-6}

   $[H^+] = 2 \times 10^{-13}$ mol dm^{-3} pH = 12.7

3 $K_a = \dfrac{[H^+][HCOO^-]}{[HCOOH]}$

4 $1.6 \times 10^{-4} = \dfrac{[H^+]^2}{0.001}$

   $[H^+]^2 = 0.001 \times 1.6 \times 10^{-4} = 1.6 \times 10^{-7}$ mol^2 dm^{-6}

   $[H^+] = \sqrt{1.6 \times 10^{-7}} = 4 \times 10^{-4}$ mol dm^{-3}

   pH = 3.4

Buffer solutions

Page 65

1 A buffer solution is a solution that maintains a constant pH despite dilution or small additions of acid or alkali. It works by having both large amounts of a proton donor and proton acceptor, which nullify small additions.

2 $K_a = \dfrac{[H^+][\text{ethanoate}]}{[\text{ethanoic acid}]}$ $1.7 \times 10^{-5} = \dfrac{[H^+]0.005}{0.001}$

   $[H^+] = \dfrac{1.7 \times 10^{-5} \times 0.001}{0.005}$

   $[H^+] = 3.4 \times 10^{-6}$ pH $= -\log_{10} 3.4 \times 10^{-6} = 5.47$

3 Chloroethanoic acid, and its salt, would be the best choice as its pK_a is closest to the desired pH. Fine tuning of the ratio of salt to acid then produces the required pH.

   Desired pH = 3.1 = $-\log[H^+]$
   $[H^+] = 7.94 \times 10^{-4}$

   Using the expression for K_a

   $1.2 \times 10^{-3} = 7.94 \times 10^{-4} \dfrac{[\text{salt}]}{[\text{acid}]}$

   Ratio [salt : acid] $= \dfrac{1.2 \times 10^{-3}}{7.94 \times 10^{-4}} = 1.5$

   i.e. the buffer needs to be made up in a ratio of 1.5 moles of salt to 1 mole of acid.

Medicines by Design (MD)

Aldehydes and ketones

Page 69

1 (a) butan-2-ol $CH_3CH_2CHOHCH_3$
   (b) pentan-1-ol $CH_3CH_2CH_2CH_2CH_2OH$
   (c) propan-1-ol $CH_3CH_2CH_2OH$

2 You would see (a) a colour change from orange to green (b) no change, stays orange.

A summary of organic reactions

Page 73

1

(a) primary amide

(b) ester

(c) acyl chloride

2 (a) $CH_3CH_2CONHC_6H_5 \xrightarrow{H^+(aq)} CH_3CH_2COOH + {}^+NH_3C_6H_5$

   (b) $C_6H_5COOCH_3 \xrightarrow{H^+(aq)} C_6H_5COOH + CH_3OH$

   (c) $CH_3COCl \xrightarrow{H^+(aq)} CH_3COOH + HCl$

3 (a) chloromethane; (b) and (c) $CH_4 \xrightarrow{Cl_2/h\nu} CH_3Cl$

   $CH_3Cl \xrightarrow[\text{reflux}]{NaOH(aq)} CH_3OH + HCl$

4 (a) 2-nitromethylbenzene; (b) and (c)

5 (a)

   (b) $CH_3CH_2CH_2NH_2 + HCl \longrightarrow CH_3CH_2CH_2NH_3^+ Cl^-$

   (c) $CH_3CH_2CH_2OH \xrightarrow{Al_2O_3(s),\ 300°C} CH_3CH{=}CH_2 + H_2O$
   or conc. H_2SO_4, reflux

Visiting the Chemical Industry (VCI)

Page 75

1 A process in which the stages of production occur at separate times in a sequence.

2 Azo dyes, because small quantities are produced in a multi-stage process.

3 (a) The reactants that go into a production process.
   (b) Hydrogen and nitrogen.
   (c) Air for nitrogen, water, and natural gas for hydrogen.

4 The infrastructure (such as road network) is already in place.

Index